NATURAL ENVIRONMENT RESEARCH COUNCIL

INSTITUTE OF TERRESTRIAL ECOLOGY

CULTURE CENTRE OF ALGAE AND PROTOZOA

LIST OF STRAINS

Edited by E. A. GEORGE

First published 1966

2nd Edition 1971

3rd Edition 1976

The Culture Centre of Algae and Protozoa is part of the Institute of Terrestrial Ecology, which itself is a component body of the Natural Environment Research Council.

CONTENTS

THE CULTURE CENTRE OF ALGAE AND PROTOZOA

36 Storey's Way, Cambridge, CB3 ODT
Telephone: Cambridge (0223) 61378

Director	E. A. George, MA
Curator of Freshwater Algae	D. J. Hibberd, PhD
Assistants	D. R. Perdue, MPhil S. F. Cann
Curator of Marine Algae	J. H. Belcher, DSc
Assistants	N. C. Pennick, BSc J. P. Cann
Curator of Protozoa	F. C. Page, PhD
Assistants	S. M. Blakey, BSc L. Greer
Head of Electron Microscopy Section	E. M. F. Swale, DSc
Technician	K. J. Clarke
Head of Preservation Section	G. J. Morris, PhD

Requests for cultures or information should be addressed to the Director.

INTRODUCTION

This collection of cultures of algae and protozoa was started by E. G. Pringsheim in Prague and the first list, comprising 48 strains, was published in 1928. It was taken over by Cambridge University in 1947 and subsequently collections at the Universities of Indiana and Göttingen were started each with a nucleus of cultures from Cambridge. Cultures left behind in Prague by Pringsheim were built into another collection there. Thus, in many cases the same strain may be in several or all of these great collections; on the other hand, in spite of considerable interchange, no one collection maintains all available worthwhile strains. In April 1970, the collection was expanded to form the basis of the Culture Centre of Algae and Protozoa, a component body of the Natural Environment Research Council. In November 1975, the Centre was merged into the Institute of Terrestrial Ecology - another and much larger component body of NERC.

As well as algae and free-living protozoa, the Cambridge collection maintains a few other micro-organisms which are not easily available elsewhere, a number of bryophytes and at the moment one flowering plant, *Wolffia*, and one rotifer. The collection does not aim to include parasitic protozoa or the larger seaweeds.

EXPLANATORY NOTES ON THE LIST

Genera, and species within them, are arranged alphabetically. Algae, protozoa, and other micro-organisms are listed together first, then the bryophytes and higher plants. A systematic list of genera is given on page 9.

In general, unless a mistake has clearly been made, or the taxonomy has been revised, names are those given to the strain by the isolator. The main objective in compiling this list has been that of enabling workers who are not well versed in taxonomy and nomenclature to find what they want with the minimum of trouble.

This list must not be taken as an authority for names. Strains are listed under the name of the species, with the following information where available:-

(a) The reference number in the collection, prefixed by 'L' if the culture is only available in liquid medium, and by 'B' if only with bacteria (agnotoxenic or monoxenic). In the absence of any such prefix, a strain may be assumed to be in pure (axenic) culture on an agar slope. 'N' indicates that the strain is preserved in liquid nitrogen (see Preservation below). An asterisk after a strain number (chiefly diatoms) indicates that the strain probably cannot be maintained indefinitely and may be replaced by a similar strain.

(b) The name of the isolator and the year of isolation. Where two names are separated by a stroke, usually the first established the agnotoxenic culture and the second rendered it axenic.

(c) Any reference number given by the isolator, or by another collection in cases where the strain was obtained from it.

(d) Any other information including an ecological note if other than fresh water, concluding with the country where the wild material was collected.

The absence of a strain or species from the list does not necessarily mean that it is not available. In some cases, we are able to obtain a strain from another collection.

Requests for organisms which lie outside the scope of CCAP should be made to other major culture collections in the United Kingdom.

OTHER CULTURE COLLECTIONS

Bacteria of industrial importance

National Collection of Industrial Bacteria,
Torry Research Station,
PO Box 31,
135 Abbey Road,
ABERDEEN,
AB9 8DG,
Scotland

Bacteria of marine importance

National Collection of Marine Bacteria,
Torry Research Station,
PO Box 31,
135 Abbey Road,
ABERDEEN,
AB9 8DG,
Scotland.

Bacteria of medical and veterinary importance

National Collection of Type Cultures,
Central Public Health Laboratory,
Colindale Avenue,
LONDON,
NW9 5HT.

Bacteria from milk and milk products

National Collection of Dairy Organisms,
National Institute for Research in Dairying,
Shinfield,
READING,
Berkshire,
RG2 9AT.

Bacteria pathogenic for plants

National Collection of Plant Pathogenic Bacteria,
Plant Pathology Laboratory,
Hatching Green,
HARPENDEN,
Hertfordshire,
AL5 2BD.

Fungi and yeasts pathogenic for man and animals

Mycological Reference Laboratory,
London School of Hygiene and Tropical Medicine,
Keppel Street (Gower Street),
LONDON,
WCIE 7HT.

Fungi, wood rotting

Building Research Establishment, Princes Risborough Laboratory,
Princes Risborough,
AYLESBURY,
Buckinghamshire,
HP17 9PX.

Fungi (other than animal pathogens and wood-rotting fungi)

Collection of Fungus Cultures,
Commonwealth Mycological Institute,
Ferry Lane,
KEW,
Surrey,
TW9 3AF.

Yeasts (other than pathogens)

National Collection of Yeast Cultures,
Brewing Industry Research Foundation,
Lyttel Hall,
NUTFIELD,
Near Redhill,
Surrey,
RH1 4HY.

STRAIN NUMBERS AND REFERENCES TO THE USE OF CULTURES

Specific names are, in effect, expressions of taxonomic opinion and as such liable to change: precise identification of strains can only be achieved by absolute fixity of strain number. Strain numbers will, therefore, never be altered whatever changes are made to specific names or groupings. If a strain is irreplaceably lost, its strain number will never be used again. Unfortunately, when the Göttingen collection was founded, the Cambridge strain numbers continued to be used there and later additions were independently given new numbers in the old notation. Hence, in some cases different strains have been given the same number. Where confusion is likely, the acronym CCAP should be used as a prefix to our strain numbers.

A sample survey of the literature on algae and protozoa showed that authors do not always give adequate reference to the strains used in work involving cultures. The sample used was the 1971 issues of fourteen journals taken at this establishment which publish original work on algae and protozoa. Papers dealing with fossils, larger seaweeds and organisms not so far cultured, were excluded.

Minimum adequate reference was considered to be the designation of the culture together with, where appropriate, indication of the source collection, e.g. CCAP 211/8d or Göttingen 11/6. A more complete reference would give also the name of the isolator and the date of isolation, but this information is usually available in the list published by the collection.

The survey showed that over three-quarters (153 out of 204) of the authors used cultures when this was possible. It also showed that well over half of the users (89 out of 153) gave inadequate or no reference to the cultures used. This is most unsatisfactory, especially when one considers the rigid insistence by authors and editors on proper bibliographic references.

References to specific names and a collection, e.g. "*Chlamydomonas globosa* from CCAP" are not satisfactory as there may be now, or in the future, more than one strain fitting that description. References such as "*Tetrahymena pyriformis* 'W' strain" are inadequate without mention of the source. It has recently been shown (Borden, Whitt and Nanney, 1973) that strains of this species from different sources, but with the same designation, may differ, while differently designated strains may be identical. The cause of this confusion presumably lies in mislabelling and failure to record the origin of stocks used.

Among the advantages of using properly documented strains are:-

(a) that the work can be repeated or compared with other work on the same strain; and

(b) the comparison of different strains of a species or of different species of a genus enables the significance of the particular characters to be assessed.

Too often one sees a general statement about a species made from evidence derived from only one strain.

"TYPE" CULTURES

When a strain is described as 'type material', it has been directly derived from material from which the original description was made. It will usually have been deposited by the author of the taxon.

PRESERVATION

The long-term preservation of viable cultures by lyophilisation or storage under liquid nitrogen is now standard practice for many kinds of micro-organisms. The main advantages are greater stability of type, less handling and hence less risk of contamination or mislabelling and less space needed for storage.

Lyophilised cultures are easily stored and transported. Cultures frozen in liquid nitrogen seem to provide much better survival for many types, but the need to maintain extremely low temperatures imposes inevitable restrictions. Progress toward routine preservation of algae and protozoa has, until now, been slow.

A programme was started in CCAP in 1974 to develop methods suitable for our strains. Despite promising results with liquid nitrogen, it will be several years before the majority of strains are successfully preserved.

REQUESTS FOR CULTURES

All cultures in this list are available for research or teaching. Strains which may be serious pathogens are only supplied to laboratories with proper facilities for handling them. All requests should be addressed to:-

The Director,
Culture Centre of Algae and Protozoa,
Institute of Terrestrial Ecology,
36 Storey's Way,
CAMBRIDGE,
CB3 ODT,
England.

It is strongly recommended that the use for which cultures are required is stated with the order, as, for example, some strains are quite unsuitable for elementary teaching and others very difficult to grow in quantity. For research purposes, if no particular strain is required, some indication of the requirement to be fulfilled will enable us to select a suitable strain.

A *month's notice* is usually necessary, especially in the case of large orders for teaching material; priority is naturally given to orders received in good time. When cultures are required on a certain date, this should be clearly indicated; otherwise they will be sent as soon as they are ready.

Cultures are dispatched by First Class Mail or by Air Mail as appropriate. Special request must be made if cultures are to be sent by Air Freight.

A charge is made to cover the cost of production, packing, etc. For details, see current price sheet.

TEACHING

List of organisms recommended for teaching purposes

Orders from this list should be headed "For teaching". Only the genus should normally be specified, as every effort is made to supply the most suitable strain available.

One culture is normally sufficient for 25 students (of *Acetabularia*, a minimum of six mature caps is supplied).

CYANOPHYCEAE

Anabaena
Chroococcus
Gloeocapsa
Gloeotrichia **(does not form typical colonies)**
Merismopedia
Nodularia **(marine)**
Nostoc
Oscillatoria
Scytonema
Spirulina
Tolypothrix

CHLOROPHYCEAE

Volvocales
Asteromonas **(marine)**
Carteria
Chlamydomonas **(for morphology)**
Chlamydomonas **(mating strains, usually available)**
Dunaliella **(marine)**
Eudorina
Gonium
Haematococcus
Pandorina
Pleodorina
Volvox

Ulotrichales
Cylindrocapsa
Microspora
Ulothrix (Uronema)

Chlorococcales
Ankistrodesmus
Chlorella
Chlorococcum
Coelastrum
Hydrodictyon
Pediastrum
Scenedesmus

Chaetophorales
Chaetophora
Coleochaete
Stigeoclonium

Cladophorales
Cladophora

Zygnemales
Closterium
Cosmarium
Micrasterias
Mougeotia
Spirogyra
Staurastrum
Zygnema

Oedogoniales
Bulbochaete
Oedogonium **(male & female strains)**

Siphonales
Acetabularia **(marine)**

PRASINOPHYCEAE

Platymonas **(marine)**
Prasinocladus **(marine)**
Pyramimonas **(marine)**

EUGLENOPHYCEAE

Astasia **(colourless)**
Euglena gracilis **(for movement)**
Euglena spirogyra **(for cell structu**
Peranema **(holozoic, shows**
Trachelomonas **flagellum well)**

DINOPHYCEAE

Amphidinium (marine)
Exuviaella (marine)
Glenodinium (marine)
Gymnodinium (marine)
Gymnodinium (Woloszynskia) (marine or freshwater)
Oxyrrhis (marine)
Peridinium (marine or freshwater)
Prorocentrum (marine)

CRYPTOPHYCEAE

Chilomonas (colourless)
Cryptomonas (marine or freshwater)

CHRYSOPHYCEAE

Ochromonas (marine or freshwater)
Pseudopedinella (marine)
Synura

PHAEOPHYCEAE

Ectocarpus (marine)

RHODOPHYCEAE

Acrochaetium (marine)
Chroothece (marine)
Porphyridium (marine) unicellular

XANTHOPHYCEAE

Botrydiopsis
Bumilleria (larger than *Tribonema* and shows H-pieces better)
Ophiocytium
Tribonema
Vaucheria
Vaucheria (fertile, limited amount of material only)

HAPTOPHYCEAE

Chrysochromulina (marine)
a coccolithophorid (marine)
Isochrysis (marine)

BACILLARIOPHYCEAE

Coscinodiscus (marine) centric
Navicula (marine) motile
Nitzschia (marine) motile
Phaeodactylum (marine)
Pinnularia
Skeletonema (marine)

SARCODINA

Amoebae:

Amoeba proteus
Chaos carolinense (larger than *A. proteus*; multinucleate)
Naegleria gruberi (amoeba/flagellate transformation; encystment and excystment)
Saccamoeba limax (medium sized; good for amoeboid movement & contractile vacuole)

Testate amoebae:

Arcella

Cellular slime moulds:

Dictyostelium (aggregation, sorocarps)

Heliozoans:

Actinosphaerium (larger than *Actinophrys*)
Actinophrys
Acanthocystis (scales of silica)

CILIOPHORA

Paramecium caudatum
Paramecium bursaria (for conjugation, zoochlorellae)
Tetrahymena (axenic)
Coleps (with 'armour')
Discophrya (suctorian)
Stentor
Spirostomum
Blepharisma
Neobursaridium (large)
Stylonychia (hypotrich)
Euplotes (hypothrich; our present species are marine)

Rotifers are also available

DEPOSITION OF CULTURES

Intending depositors should first apply to the Senior Officer.

We generally welcome the deposition of cultures of value whether of taxonomic, physiological or other importance, especially those that have been the subject of publication. It is recommended that important strains should be deposited in at least two major collections for safe keeping.

Depositors are requested to supply as much information as possible about their strains, preferably on our data sheets (available on request), or on World Federation of Culture Collections form SCC-4. Reprints relating to strains are most useful.

If desired, a newly-deposited culture may be withheld from issue for a limited time, usually until publication is effected.

Depositors may, within limits, receive cultures of their own strains free of charge.

Arrangements can be made for the deposition of strains subject to patent applications.

SYSTEMATIC LIST OF GENERA OF ALGAE AND PROTOZOA IN THE COLLECTION

CHLOROPHYCEAE

Volvocales

Balticola
Brachiomonas
Carteria
Chlamydomonas
Chlorogonium
Dunaliella
Dysmorphococcus
Eudorina
Gloeomonas
Gonium
Haematococcus
Lobomonas
Pandorina
Pascherina
Phacotus
Platydorina
Pleodorina
Polytoma
Polytomella
Pteromonas
Pyrobotrys
Spermatozopsis
Sphaerellopsis
Stephanosphaera
Volvox
Volvulina

Chlorococcales

Ankistrodesmus
Asterococcus
Borodinella
Bracteacoccus
Characium
Chlorella
Chlorochytrium
Chlorococcum
Chlorothecium
Coccomyxa
Coelastrella
Coelastrum
Crucigenia
Cystococcus
Dactylococcus
Dictyococcus
Dictyochloris
Dictyosphaerium
Dimorphococcus
Elakotothrix
Eremosphaera
Fusola
*Glaucocystis**
Golenkiniopsis
Gomontia
Halochlorella
Halochlorococcum
Heterogonium
Hyaloraphidium
Hydrodictyon
Hypnomonas
Kentrosphaera
Kirchneriella
Lagerheimia
Micractinium
Muriella
Myrmecia
Nannochloris
Neospongiococcum
Nephrochlamys
Oocystis
Pediastrum
Polyedriopsis
Prototheca
Pseudochlorococcum
Pseudococcomyxa
Quadrigula
Radiosphaera
Rhopalocystis
Scenedesmus
Scotiella
Selenastrum
Spongiochloris
Tetraedron
Trebouxia

* **Robinson and Preston (1971) suggest that *Glaucocystis* could be: (1) a red alga; (2) a primitive dinoflagellate; or (3) of an independent group; but not a blue-green symbiont in a colourless green alga.**

Tetrasporales

Actinochloris
Asterococcus
Chaetopeltis
Glaucosphaera
Gloeocystis
Nautococcus
Neochloris
Palmodictyon
Paulschulzia
Tetracystis

Ulotrichales

Cylindrocapsa
Interfilum
Klebsormidium (Hormidium)
Microspora
Prasiococcus
Prasiola
Pseudostichococcus
Schizomeris
Sphaeroplea
Stichococcus
Ulothrix
Uronema

Chaetophorales

Caespitella
Cephaleuros
Chaetophora
Chlorosarcinopsis
Chlorosphaera
Coleochaete
Dilabifilum
Diplosphaera
Draparnaldia
Gongrosira
Leptosira
Microthamnion
Physolinum
Pilinia
Pleurastrum
Pleurococcus
Pseudendocloniopsis
Pseudendoclonium
Pseudopleurococcus
Raphidonema
Stigeoclonium
Trentepohlia

Cladophorales

Cladophora
Pithophora
Rhizoclonium

Conjugales

Closterium
Cosmarium
Cylindrocystis
Euastrum
Gonatozygon
Hyalotheca
Mesotaenium
Micrasterias
Mougeotia
Netrium
Pleurotaenium
Sphaerozosma
Spirogyra
Spondylosium
Staurastrum
Zygnema
Zygnemopsis

Oedogoniales

Bulbochaete
Oedocladium
Oedogonium

Siphonales (Bryopsidales)

Acetabularia
Derbesia
Dichotomosiphon
Protosiphon

PRASINOPHYCEAE

Asteromonas
Bipedinomonas
Halosphaera
Heteromastix
Mesostigma
Micromonas
Monomastix
Nephroselmis
Pedinomonas
Platymonas
Prasinocladus
Pyramimonas
Stephanoptera
Tetraselmis

XANTHOPHYCEAE

Botrydiopsis
Botrydium
Botryococcus
Bumilleria
Bumilleriopsis
Chlorellidium
Chloridella
Chlorocloster
Heterococcus
Heterothrix
Mischococcus
Nephrodiella
Ophiocytium
Pleurochloris
Sphaerosorus
Tribonema
Vaucheria

EUSTIGMATOPHYCEAE

Chlorobotrys
Ellipsoidion
Monodus
Pleurochloris
Polyedriella
Pseudocharaciopsis
Vischeria

CHRYSOPHYCEAE

Anthophysa
Chrysococcus
Chrysosphaera
Mallomonas
Monas
Ochromonas
Olisthodiscus
Paraphysomonas
Phaeaster
Poterioochromonas
Pseudopedinella
Spumella
Synura

HAPTOPHYCEAE

Chrysochromulina
Coccolithus
Cricosphaera
Crystallolithus
Diacronema
Dicrateria
Emiliania
Hymenomonas
Isochrysis
Monochrysis
Pavlova
Phaeocystis
Pleurochrysis
Prymnesium
Sarcinochrysis
Syracosphaera

DINOPHYCEAE

Amphidinium
Exuviaella (Prorocentrum)
Gloeodinium
Heterocapsa
Oxyrrhis
Peridinium
Prorocentrum
Woloszynskia

BACILLARIOPHYCEAE

Amphiprora
Asterionella
Bacillaria
Biddulphia
Chaetoceros
Cyclotella
Fragilaria
Gomphonema
Melosira
Navicula
Nitzschia
Pinnularia
Pleurosigma
Skeletonema
Stephanodiscus
Tabellaria
Thalassiosira

CRYPTOPHYCEAE

Chilomonas
Chroomonas
Cryptomonas
Cyanophora
Cyathomonas
Hemiselmis

CHLOROMONADOPHYCEAE

Fibrocapsa
Vacuolaria

EUGLENOPHYCEAE

Astasia
Colacium
Cryptoglena
Cyclidiopsis
Distigma
Entosiphon
Euglena
Eutreptia
Eutreptiella
Gyropaigne
Hyalophacus
Khawkinea
Lepocinclis
Menoidium
Parmidium
Peranema
Phacus
Rhabdomonas
Rhabdospira
Strombomonas
Trachelomonas

PHAEOPHYCEAE

Ectocarpus
Streblonema

RHODOPHYCEAE

Asterocytis
Chantransia
Chroothece
*Cyanidium***
*Glaucocystis**
Hildenbrandia
Porphyra
Porphyridium
Rhodella

** **This alga, which at first glance is a 'blue-green *Chlorella*', has been placed in various other classes**

* **see footnote on page 9**

MYXOPHYCEAE (= Cyanophyceae)

Anabaena	*Lyngbya*
Anabaenopsis	*Mastigocladus*
Anacystis	*Merismopedia*
Aphanizomenon	*Microchaete*
Aphanocapsa	*Microcystis*
Arthrospira	*Myxosarcina*
Calothrix	*Nodularia*
Chlorogloea	*Nostoc*
Chroococcus	*Oscillatoria*
Chroococcopsis	*Pelogloea*
Coccochloris	*Phormidium*
Coelosphaerium	*Plectonema*
Cylindrospermum	*Pseudanabaena*
Dermocarpa	*Pseudoholopedia*
Entophysalis	*Schizothrix*
Fischerella	*Scytonema*
Fremyella	*Spirulina*
Gloeocapsa	*Symploca*
Gloeotrichia	*Synechococcus*
Lampropedia	*Synechocystis*
Lauterbornia	*Tolypothrix*

ZOOMASTIGOPHOREA

Bodo	*Rhynchomonas*

SARCODINA

Amoebida

Acanthamoeba	*Naegleria*
Adelphamoeba	*Paramoeba*
Amoeba	*(Pelomyxa;* **see** *Chaos)*
Cashia	*Platyamoeba*
Chaos	*Rosculus*
Echinamoeba	*Saccamoeba*
Filamoeba	*Stachyamoeba*
Flabellula	*Tetramitus*
Hartmannella **(see also** *Acanthamoeba, Cashia, Echinamoeba, Saccamoeba)*	*Thecamoeba*
	Vahlkampfia
Heteramoeba	*Vannella*
Mayorella **(see also** *Acanthamoeba***)**	*Vexillifera*

Arcellinida (lobose testaceans)

Arcella	*Cryptodifflugia*
Cochliopodium	

Gromiida (filose testaceans)

Euglypha	*Trinema*

Leptomyxida

Leptomyxa

Mycetoxoa

Dictyostelium

Heliozoia

Acanthocystis
Actinophrys
Actinosphaerium (see also *Echinosphaerium*)
Echinosphaerium
Raphidiophrys

CILIOPHORA

Primociliatida

Coleps

Colpodida

Colpoda

Suctorida

Discophrya
(*Podophrya*; see *Discophrya*)

Hymenostomatida

Colpidium
Paramecium
Tetrahymena
Uronema

Scuticociliatida

Cyclidium

Peritrichida

Vorticella

Heterotrichida

Blepharisma
Neobursaridium
Spirostomum
Stentor

Hypotrichida

Euplotes
Stylonychia

ROTIFERA

Bdelloidea

Philodina

CULTURE MEDIA

Media may be designed for maintenance, i.e, giving prolonged survival between subculturing, for maximum growth rate, for inducing sexuality, or other special purposes. They may be chemically defined or loosely defined, i.e. containing imprecise ingredients such as soil or extracts of yeast, meat, etc. When cultures have other micro-organisms such as bacteria present, the medium must not be such as to encourage excess growth of contaminants lest they overwhelm the desired organism. In general, media with a total of 0.1 per cent or more organic nutrient are only suitable for axenic cultures.

Large numbers of cultures are maintained on the same medium mainly for convenience. It must not be assumed that one medium is optimal for many strains. Frequently, related species, or even strains within a single species, have significantly different requirements.

Details of some of our more useful maintenance media are given below. Many special media and methods for algae are given by Stein (1973) and Venkataraman (1969).

Proteose peptone (Difco)	0.1%*
KNO_3	0.02%
K_2HPO_4	0.002%
$MgSO_4.7H_2O$	0.002%
Agar	1.0%

This medium proves satisfactory for many algae and also quickly reveals the presence of many common contaminants. For agar cultures with bacteria, a less rich medium is necessary; we generally use soil extract agar (e + s).

Soil extract stock solution	10% by volume
KNO_3	0.02%
K_2HPO_4	0.002%
$MgSO_4.7H_2O$	0.002%
Agar	1.0%

The soil extract stock solution is made by heating in a steamer a calcareous garden loam with twice its volume of supernatant water.

It is convenient to make up and sterilise a number of small containers of stock solution each of a size appropriate to making a batch of media. Repeated autoclaving is deleterious; excessive bacterial growth would spoil the stock solution.

Soil and water media

In spite of possible initial difficulties over selection of suitable ingredients, these media have great advantages for many purposes so long as axenic culture is not required. A well-chosen soil and water medium produces a morphologically excellent growth of almost any organism apart perhaps from the more exacting planktonic forms. The soil

* Unless otherwise indicated % means g per 100 ml.

provides not only mineral nutrients and trace elements, but the latter are naturally chelated, pH is buffered, and toxic products rendered innocuous. The microbial flora ensures a supply of growth substances.

Our basic medium "E" is made as follows:-

Put a layer about 1 cm deep of good calcareous garden loam into a test tube or jar. Carefully add 7 to 10 cm of water, plug or cover, and steam for one hour (longer for larger vessels) on each of two consecutive days. Allow to stand a further day before inoculating, when the pH should be between 7 and 8. The use of mud from rivers or ponds is seldom as satisfactory as garden soil.

Innumerable variations of this basic medium are possible. Among those we find useful are "M" in which the garden soil is replaced by a calcareous clay. The addition of a speck (say about 3% of the volume of the soil) of calcium carbonate or of ammonium magnesium phosphate beneath the soil is recommended, the former for many eutrophic Chlorophyceae, the latter for many green euglenids. Sphagnum peat may be added or may replace the soil when growing forms from acid habitats. The addition of a little starch below the soil stimulates growth of many saprophytes like *Polytoma* and *Astasia*. A pearl barley grain, rice grain or wheat grain produces a bacterial flora which forms a suitable food for many ciliates. By giving consideration to the nutrition of the organism and the nature of its habitat, one can produce soil and water media to suit many different organisms.

When selecting soils, the best method is to take a fair-sized sample, say about one cubic foot, pass this through a sieve of about 1 cm mesh and then make up media and test them with a range of organisms. Should a sample prove satisfactory, you have a sufficient stock, it being often difficult to collect an identical sample even from the same locality.

It is strongly recommended that, where possible, teachers obtain inocula from the collection well in advance and grow them in their own laboratory, using the soil and water method described above. This has the advantage of introducing the students to a culture method based on ecological principles, and with a little practice the material will be in perfect condition. With many algae, e.g. *Euglena*, *Oedogonium* and *Ochromonas*, the cell contents often become obscured by accumulated food reserves in old cultures. In other forms, e.g. *Trachelomonas* and *Pithophora*, certain morphological features only become apparent in older cultures.

<u>Bold's basal medium</u>, (Bischoff and Bold, 1963), is a very useful mineral medium for a wide range of algae including those from more or less eutrophic waters or from soils. It may also be supplemented by soil extract and/or vitamins.

Six stock solutions 400 ml in volume are employed, each containing one of the following salts in the concentration listed:-

$NaNO_3$	10.0g
$CaCl_2.2H_2O$	1.0g
$MgSO_4.7H_2O$	3.0g
K_2HPO_4	3.0g
KH_2PO_4	7.0g
NaCl	1.0g

To 940 ml of distilled water are added 10 ml of each stock solution and 1.0 ml of each of the four stock trace-element solutions prepared as follows:-

1. 50 g EDTA and 31 g KOH dissolved in 1 litre H_2O.

2. 4.98 g $FeSO_4.7H_2O$ dissolved in 1 litre acidified H_2O. (Acidified H_2O: 1.0 ml H_2SO_4 added to 999 ml distilled water.)

3. 11.42 g H_3BO_3 dissolved in 1 litre H_2O.

4. The following, in amounts indicated, all dissolved in 1 litre H_2O: $ZnSO_4.7H_2O$, 8.82 g; $MnCl_2 4H_2O$, 1.44 g; MoO_3, 0.71 g; $CuSO_4.5H_2O$, 1.57 g; $Co(No_3)_2.6H_2O$, 0.49 g.

Chu (1942) devised a series of media based on the composition of various natural waters. Chu 10, variously modified, has been used for a wide range of green and blue-green algae and also diatoms, particularly from oligotrophic or less rich eutrophic habitats.

The following is given by Gerloff, Fitzgerald and Skoog (1950).

$Ca(NO_3)_2$	0.004%
K_2HPO_4	0.001%
$MgSO_4.7H_2O$	0.0025%
Na_2CO_3	0.002%
Na_2SiO_3	0.0025%
Ferric citrate	0.0003%
Citric acid	0.0003%

Nowadays, the last two ingredients are usually replaced by an iron/EDTA complex, and other trace elements and vitamins may be added as in this modification from the Windermere Laboratory of the Freshwater Biological Association.

<u>Modified Chu 10 (EVT)</u>

		amounts per litre
	$Ca(NO_3)_2.4H_2O$	20.0 mg
	KH_2PO_4	6.2 mg
	$MgSO_4.7H_2O$	25.0 mg
	Na_2CO_3	20.0 mg
	Na_2SiO_3	25.0 mg
	N.HCl	0.25 ml
Added as complex EDTA.Fe	$EDTA.Na_2$	2.0 mg
	$FeCl_3$	1.0 mg
	H_3BO_3	2.48 mg
	$MnCl_2.4H_2O$	1.39 mg
	$(NH_4)_6Mo_7O_{24}.4H_2O$	1.00 mg
	vitamin B_{12}	0.25 mg
	vitamin B_1	1.00 mg
	biotin	1.00 mg

Many strains require more specialized media; some of the most important are these:-

Euglena gracilis medium (E.g.)

Na acetate. hydrate	0.1%
Beef extract	0.1%
Yeast extract	0.2%
Bacto tryptone	0.2%
$CaCl_2$	0.001%
Agar (optional)	1.0%

Ochromonas medium (Ochr)

Liver infusion (dehydrated) Oxoid	0.1%
Glucose	0.1%
Bacto tryptone	0.1%

Tetrahymena medium (P.Y.)

Proteose peptone	1.0%
Yeast extract	0.25%

Polytoma medium (Polyt.)

Na acetate. hydrate	0.2%
Yeast extract	0.1%
Bacto tryptone	0.1%
Agar (optional)	1.0%

Acanthamoeba, when axenic, is maintained in 4 per cent mycological peptone (M.P.).

Marine algae

Many marine algae thrive in Erdschreiber medium:-

Natural seawater	1 litre
Soil extract stock solution	50 ml
$NaNO_3$	0.2 g
$Na_2HPO_4.12H_2O$	0.03 g

The seawater is filtered, pasteurized, and added to the other ingredients after they have been autoclaved.

For more exacting marine algae, media such as ASP2 (Provasoli et al, 1957) are used, often mixed with an equal part of Erdschreiber. Organisms from polluted marine or estuarine habitats often thrive in a soil and water medium made with an appropriate concentration of seawater.

The following are only two of many liquid media used for *Chlorella*, *Chlamydomonas* and other green algae.

Chlorella, Sorokin and Krauss (1958)

	(g/litre)
KNO_3	1.25
KH_2PO_4	1.25
$MgSO_4.7H_2O$	1.00
$CaCl_2$	0.0835
H_3BO_3	0.1142
$FeSO_4.7H_2O$	0.0498
$ZnSO_4.7H_2O$	0.0882
$MnCl_2.4H_2O$	0.0144
MoO_3	0.0071
$CuSO_4.5H_2O$	0.0157
$Co(NO_3)_2.6H_2O$	0.0049
EDTA	0.5

Chlamydomonas reinhardii

For genetic studies, this alga has been grown in the following medium by Sueoka (1960) and Levine's group at Harvard:-

	Minimal medium
* NH_4Cl	0.05 g
$MgSO_4.7H_2O$	0.02 g
$CaCl_2.2H_2O$	0.01 g
** K_2HPO_4	0.72 g
** KH_2PO_4	0.36 g
Hutner's trace elements	1 ml
Distilled water	1 litre

Hutner's trace elements solution:	(g)
EDTA	50.0
$ZnSO_4.7H_2O$	22.0
H_3BO_3	11.4
$MnCl_2.4H_2O$	5.1
$FeSO_4.7H_2O$	5.0
$CoCl_2.6H_2O$	1.6
$CuSO_4.5H_2O$	1.6
$(NH_4)_6Mo_7O_{24}.4H_2O$	1.1
Distilled water (ml)	750

Boil, cool slightly, and bring to pH 6.5 - 6.8 with KOH (do not use NaOH). The clear solution is diluted to 1000 ml with distilled water and should have a green colour which changes to purple on standing. It is stable for at least one year.

For heterotrophic acetate mutants, the medium may be supplemented with sodium acetate at a concentration in the medium of 0.2%.

* concentration may be increased 10-fold

** concentration may be doubled

Blue-green algae

Medium C of Kratz and Myers (1955) is a favourite medium used by investigators working with *"Anacystis nidulans"* or other blue-green algae.

	(g/litre)
$MgSO_4.7H_2O$	0.25
K_2HPO_4	1.00
$Ca(NO_3)_2.4H_2O$	0.025
KNO_3	1.0
Na_3 citrate.$2H_2O$	0.165
$Fe_2(SO_4)_3.6H_2O$	0.004
A_5 microelements	1.0 ml

Microelements stock solution:	
H_3BO_3	2.86
$MnCl_2.4H_2O$	1.81
$ZnSO_4.7H_2O$	0.222
MoO_3 (85%)	0.0177
$CuSO_4.5H_2O$	0.079

Several other media for blue-green algae are given by Carr and Whitton (1973, Appendix B).

Media for protozoa

Free-living protozoa, apart from the phytoflagellates are nearly all facultative or obligate phagotrophs. Only a few forms such as *Tetrahymena* and *Acanthamoeba* thrive in axenic liquid culture. Most protozoa grow best in agnotoxenic culture with bacteria and perhaps other micro-organisms as food. In some cases, they can be grown in monoxenic culture with one known food organism.

Free swimming ciliates such as *Paramecium* and *Spirostomum* thrive in soil and water medium (see page 15) with the addition of a barley grain or other source of nutrient to provide a flourishing bacterial flora as food for the ciliate. Other protozoa may be more exacting and, in any case, sessile or bottom-living forms may be difficult to separate from the soil particles. Details of media are available on request.

A simple method for the cultivation of *Amoeba proteus*

This method, not original with us, has proved very satisfactory for the easy maintenance of cultures for teaching purposes, assuming that the inoculum includes not only *Amoeba proteus*, but also a food organism (e.g. *Chilomonas paramecium*).

Pour Prescott's and James's solution (see below) into a dish such as a crystalizing dish (diameter ca. 100 mm or a little smaller) to a depth of approximately 1 cm. Add 3 or 4 uncooked rice grains, and then inoculate with 1 or 2 ml of healthy culture of *Amoeba proteus* containing the food organisms. (If you intend to propagate the culture which you obtain from us, request a culture already containing *Chilomonas*, since we also supply from mass cultures not containing *Chilomonas*.) Cover with half a petri dish or other loose cover, and keep preferably at a temperature of 18 or 19°.

In six or eight weeks, the culture should contain good numbers of amoebae, including some among the fungal mycelia usually growing out from the rice grains, and should still be usable after three months. In a health culture, amoebae are moving with pseudopodia well extended. When most amoebae consist of only monopodial elongated forms, the culture is no longer in very good condition, and rounded cells are also usually unhealthy. Be sure to distinguish between amoebae and elongated bits of detritus.

Amoeba proteus seems to need a few bacteria, so sterilisation of the medium is not recommended. Subculture every two months.

Prescott's and James's solution

Make up three stock solutions:

A	$CaCl_2$	0.327 g
	KCl	0.162 g
	Distilled water to 100 ml	
B	K_2HPO_4	0.512 g
	Distilled water to 100 ml	
C	$MgSO_4.7H_2O$	0.280 g
	Distilled water to 100 ml	

Final solution: 1 ml each of A, B, and C in 997 ml distilled water.

If you cannot prepare the Prescott's and James's solution, try filtered and boiled water from a fairly clean pond or stream.

LIST OF ALGAE AND PROTOZOA

ACANTHAMOEBA Volkonsky

A. **astronyxis** (Ray *et* Hayes)
1534/1 Harrison 1957: material from Ray: Washington USA

A. **castellanii** (Douglas)
1501/1a Neff 1957: soil: California USA
1501/1b Neff 1957: Korn's substrain
1501/2a Castellani 1930: England
1501/2b Chang 1959: pathogenic*: Ohio USA
B 1501/2g Nagington 1974: pathogenic*: human cornea: England
1534/2 Lewin 1951: Connecticut USA
1534/3 Singh 1952, as *Hartmanella rhysodes*: soil: England

A. **comandoni** Pussard
B 1501/5 Comandon, AIP: soil: type material

A. **culbertsoni** (Singh *et* Das)
B 1501/6 Culbertson 1959 as AI: type material: pathogenic*: monkey kidney tissue culture: USA

A. **griffini** Sawyer
1501/4 Griffin 1962, No. 5-7: type material: reportedly euryhaline: Connecticut USA

A. **palestinensis** (Reich)
1547/1 Reich 1933: type material of *Mayorella palestinensis*: soil: Israel

A. **polyphaga** (Puschkarew)
1501/3a Page 1964, No. 23: Wisconsin USA
1501/3b Page 1965, No. 45: Alabama USA
1501/3c Sawyer 1967, OX-1: Maryland USA
B 1501/3d Nagington 1974: pathogenic*: human cornea: England
1501/3g Jones 1974, 'Garcia': human pathogen*: Texas USA
1501/3h Wang 1959: pathogenic*: New York USA

ACANTHOCYSTIS Carter

A. **erinaceoides** Petersen *et* Hansen
LB 1504/1 Ockleford 1970: Scotland

ACETABULARIA Lamouroux

A. **acetabulum** (L.) Silva (= *A. mediterranea* Lam.)
LB 702/1 Brachet: marine

ACROCHAETIUM Naegeli

A. **sagraeanum** (Mont.) Bornet
LB 1350/1 Ott, No. MO 366: marine

* These pathogenic strains will only be issued to recognised laboratories capable of handling them safely.

ACTINOCHLORIS Korshikov

A. sphaerica Korschikov
B 247/1 Starr 1951: from soil: England

ACTINOPHRYS Ehrenberg

A. sol Ehrenberg
LB 1502/2 Page 1973: England

ACTINOSPHAERIUM Stein

A. eichhorni (Ehrenberg)
LB 1507/2 Goodfellow 1974: England

A. nucleofilum Barrett see *Echinosphaerium*

ADELPHAMOEBA Napolitano, Wall *et* Ganz

A. galeacystis (Napolitano)
B 1506/1 Napolitano 1967: soil: type material: New York USA

AMOEBA Bory

A. proteus Leidy
LB 1503/2 Taylor as *A. lescherae*
LB 1503/3 Ward's strain
LB 1503/4 Carlsberg "A": reportedly free of cytoplasmic bacteria

AMPHIDINIUM Claparède *et* Lachmann

A. carterae Hulburt
LB 1102/1 Parke 1954: Plymouth 127: marine: England
LB 1102/2 Butcher: marine:England

A. klebsii Kofoid *et* Swezy
LB 1102/3 Butcher: marine: England

A. spp.
LB 1102/4 Butcher 1956: brackish: England
LB 1102/5 Butcher 1960: marine: England

AMPHIPRORA Ehrenberg

A. hyalina Eulenstein *ex* Van Heurck
LB 1003/1 Grell 1956: marine: a good food organism: Germany

ANABAENA (Bory) Bornet *et* Flahault

A. aequalis Borge

B	1403/2b	derived from 1403/2a after treatment with bromouracil and penicillin. Identified as *A. aequalis* by Kantz and Bold (1969)

A. ambigua Rao

LB	1403/7	Mitra: identified as *A. oscillarioides* by H. Forest: identified as *A. sphaerica* by Kantz and Bold: India

A. catenula (Kütz.) Bornet *et* Flahault

	1403/1	Manten: Utrecht P 36: identified as *A. inaequalis* by H. Forest: soil: Holland

A. cylindrica Lemmermann

	1403/2a	Chu 1939/Fogg 1940: nitrogen fixer: identity confirmed by Kantz and Bold: identified as *A. inaequalis* by H. Forest: England
B	1403/2b	derived from 1403/2a after treatment with bromouracil and penicillin: identified as *A. aequalis* by Kantz and Bold

A. flos-aquae (Lyng.) Bréb.

	1403/13a	Tischer, Univ. Indiana No. 1444
LB	1403/13b	Lund 1964: England

A. inequalis (Kütz.) Bornet *et* Flahault

	1403/1	as *A. catenula q.v.*
	1403/2a	as *A. cylindrica q.v.*
B	1403/2b	as *A. cylindrica q.v.*
	1446/1a	Utrecht P31
B	1446/1c	Utrecht P32

A. oscillarioides (Bory) Bornet *et* Flahault

LB	1403/7	as *A. ambigua q.v.*
	1403/9	Murray 1963: soil: New York USA: identity doubtful
	1403/10	Drouet 1954/Forest: Minnesota USA
	1403/11	de Halperin 1959

A. sphaerica Bornet *et* Flahault

	1403/7	as *A. ambigua q.v.*

A. subcylindrica Borge

	1403/4b	as *A. variabilis q.v.*
	1403/12	as *A. variabilis q.v.*

A. variabilis (Kütz.) Bornet *et* Flahault

	1403/4b	Utrecht P40: identified as *A. subcylindrica* by Kantz and Bold
LB	1403/8	Probably isolated by Pringsheim in Cambridge: grows well and is used for demonstration as *'Anabaena'*
	1403/12	Isolated at Bot. Inst. Halle: Griefswald A92: readily develops spores: identified as *A. subcylindrica* by Kantz and Bold

A. spp.

	1403/14	Atkinson 1964/65: Malaya
B	1403/17	Rothamsted 31A: Wilcox Strain III
B	1403/15	Wilcox 1971, Strain I: England
B	1403/16	Wilcox 1971, Strain II: England

ANABAENOPSIS Woloszyńska

A. circularis (G.S. West) Wolosz *et* Miller
1402/1 Watanabe, M4: Japan

ANACYSTIS Meneghini

A. cyanea (Kutz.) Drouet *et* Daily
LB 1450/1 as *Microcystis aeruginosa q.v.*

A. dimidiata (Kutz.) Drouet *et* Daily
LB 1412/1b as *Chroococcus turgidus q.v.*

A. marina (Hansg.) Drouet *et* Daily
LB 1479/1a as *Synechococcus elongatus q.v.*

A. montana (Lightf.) Drouet *et* Daily
LB 1404/1 as *Aphanocapsa rivularis q.v.*
1430/1 as *Gloeocapsa alpicola q.v.*

A. montana *f* minor Drouet *et* Daily
LB 1404/2 as *Aphanocapsa sp. q.v.*
LB 1405/3 Provasoli

'**A. nidulans** (Richt.) Drouet
The Kratz/Allen strain widely known under this name is certainly not *Anacystis sp.* We have listed it as *Synechococcus leopoliensis* (Racib) Komarek. See Komarek 1970.

ANKISTRODESMUS Corda, including *Monoraphidium* Fott *et* Novakova

There is uncertainty about the nomenclature and identification of some of these algae.

A. acicularis (A.Br.) Korshikov
202/11a George 1951: England
202/11b George 1951: England
202/11c Lewin 1951: Connecticut USA
202/11d Starr: Indiana USA
202/11e Wurtz, No. 14: France
202/11f George 1950: Denmark

A. angustus Bernard *(A. falcatus var. spirilliformis G. S. West)*
202/2 Vischer 1923, No. 9: Geneva No. 227: Switzerland
202/3a Rodhe, No. 1631a: Sweden
202/3b Wurtz, No. 13: France
202/3c Chodat: Prague, No. 75
202/4a Czurda c. 1929: Czechoslovakia
202/4b George 1950: England
202/4c Wurtz, No. 15: France
202/4d George 1951: England
202/4e George 1951: Finland
202/4f George 1954: Nigeria
202/4g Ross 1952: Lewin NO. 226: Connecticut USA

A. braunii Brunnthaler
202/7a Vischer 1933, No. 106: Switzerland
202/7b George 1948: South Africa
202/7c Norris 1954: California USA
202/7d George 1951: England
202/7e George 1953: England
202/7f Wurtz, No. 12: France
202/8a Vischer, No. 339
202/8b George 1950: England
202/8c Pirson 1952: Germany
202/8d Fox 1953: Nigeria
202/9 George 1950: England

A. convolutus Corda
202/10a George 1952: England
202/10b George 1952: Wales
202/10c George 1952: England
202/10d Wurtz 1947, No. 11: France
202/10e Lewin 1950: Connecticut USA
202/10f George 1951: France
202/10g George 1950: Sweden
202/10h Weis 1952, Lewin No. 225: Connecticut USA
202/10j Golterman 1958: Holland

A. cucumiformis Belcher *et* Swale
202/22 Belcher 1961: type material: England

A. curvulus Belcher *et* Swale
202/16 Belcher 1958: type material: England

A. densus Korshikov
202/1 Vischer 1923: type strain of *A. amalloides nom. nud.*: Switzerland
202/20 Belcher 1961: England

A. falcatus (Corda) Ralfs
202/5a Czurda 1942: Czechoslovakia
202/5c Algéus 1942: Sweden
202/5d George 1955: England
202/14a George 1951: France
202/14b Wurtz 1947, No. 17: France
202/14c Christensen: Bornholm
202/15a Golterman 1958: Holland
202/15b Pringsheim 1955: Germany

A. falcatus *var.* **terrestris** Bristol
202/23 Flint/King 1966: New Zealand

A. longissimus (Lemm.) Wille
202/13 Myers, No. 9: Texas USA

A. lunulatus Belcher *et* Swale
202/17 Belcher 1960: type material: England

A. nannoselene Skuja
202/6a Rodhe pre 1948, No. 1632: type material: Sweden
202/6b George 1951: England

A. pseudobraunii Belcher *et* Swale
202/19 Belcher 1962: type material: England

A. sabrinensis Belcher *et* Swale
202/21 Belcher 1961: type material: England

A. spiralis (Turner) Lemmermann
202/12 Christensen 1948, No. 4879: Bornholm

A. spirilliformis see *A. angustus*

A. subcapitatus Korshikov
202/18 Belcher 1960: England

A. sp.
LB 202/24 Butcher: marine: England

ANTHOPHYSA Bory

A. vegetans (OFM) Stein
LB 905/1 Pringsheim 1950: England

APEDINELLA Throndsen

A. spinifera (Thr.) Throndsen
LB 906/1 Throndsen 1965: Plymouth 460: type material: marine: Norway

APHANIZOMENON Morren

A. flos-aquae (L.) B. *et* F.
LB 1401/1 Heaney 1968: Ireland

APHANOCAPSA Naegeli

A. rivularis (Carm.) Rabenhorst (*Anacystis montana* (Lightf.) Drouet *et* Daily)
LB 1404/1 Pringsheim 1947: England

A. sp. (*Anacystis montana f. minor* Drouet *et* Daily)
LB 1404/2 Pringsheim 1947: England

ARCELLA Ehrenberg

A. polypora Penard
LB 1505/2 Page 1974: England

ARTHROSPIRA Stitzenberger, see also *Spirulina*

A. sp.
LB 1475/5 Lack 1971, L235: England

ASTASIA Dujardin

A. **applanata** Pringsheim
LB 1204/1 Pringsheim 1936: type strain: Austria

A. **curvata** Klebs
LB 1204/5 Pringsheim 1940: England

A. **dangeardii** Lemmermann *var.* **parva** Pringsheim
LB 1204/7 Pringsheim: type strain: Brazil

A. **fritschii** Pringsheim *(Khawkinea)*
LB 1204/8a Pringsheim 1940: England
LB 1204/8b Pringsheim 1940: England

A. **hallii** (Jahn et McKibben) Pringsheim *(Khawkinea)*
LB 1204/12 Pringsheim 1940: England

A. **inflata** Klebs
LB 1204/14 Pringsheim: Czechoslovakia

A. **klebsii** Lemmermann
LB 1204/15 Pringsheim

A. **linealis** Pringsheim *(Cyclidiopsis acus)*
LB 1204/16b Christen: Clone 1-1
LB 1204/16c Christen: Clone 3-1
LB 1204/16d Christen: Clone 4-1

A. **longa** Pringsheim
LB 1204/17a Pringsheim: type strain: Czechoslovakia
1204/17b Hall
1204/17c Pringsheim: England
1204/17d Pringsheim: arose from *Euglena* 1224/5g: Scotland
1204/17e Lackey: arose from *Euglena* 1224/5h
1204/17f Provasoli: A8: Italy
1204/17g Dach 1940: as *A. klebsii*
1204/17j Provasoli as *A. lombardica*: Italy

A. **longa** Pringsh. *var.* **truncata** Pringsheim
1204/18 Pringsheim 1938: type strain: Czechoslovakia

A. **ocellata** Khawkine *var.* **provasolii** Pringsheim *(Khawkinea)*
1204/9 Pringsheim/Provasoli 1947: type strain: Czechoslovakia

A. **pertyi** Pringsheim *(Khawkinea)*
1204/3 Pringsheim 1939: England

A. **quartana** (Moroff) Pringsheim *(Khawkinea)*
LB 1204/20a Pringsheim 1939
LB 1204/20b Pringsheim 1940
1204/17h Provasoli as *A. gambarone*: Italy

A. **sagittifera** Skuja
LB 1204/24 Christen

A. **solea** Pringsheim *(Khawkinea)*
LB 1204/19 Pringsheim 1947: England

A. torta Pringsheim
LB 1204/21 Pringsheim 1936: type strain: Austria

A. spp.
LB 1204/22 Starr 1950: England
LB 1204/23 Pringsheim: England = ? *A. margaritifera*

ASTERIONELLA Hassall

A. formosa Hassall
LB 1005/1b Jaworski 1974, L272: England

ASTEROCOCCUS Scherffel

A. siderogloeus (Pascher *et* Jahoda) Novakova
LB 31/2 George 1950: England; typical colonies not produced in culture

A. superbus (Cienk.) Scherffel
3/3a George 1949: France
3/3b George 1950: England

ASTEROCYTIS Gobi

A. ornata (Agardh) Hamel, see *Chroothece richterianum*

A. ramosa (Thwaites) Gobi
LB 1353/3 Belcher 1974: England

ASTEROMONAS Artari

A. gracilis Artari
LB 80/1 Lewin 1955: marine: California USA

A. propulsum Butcher, see *Pyramimonas aff. orientalis* Butcher

BACILLARIA Gmelin

B. paradoxa Gmelin
LB 1006/1* Belcher 1974: marine: England

BALTICOLA Droop (*Haematococcus* Agardh *q.v.*)

BIDDULPHIA Gray

B. sp.
LB 1007/1* Belcher 1974: marine

BIPEDINOMONAS Carter, see *Heteromastix*

BLEPHARISMA Perty

B. americanum Suzuki
LB 1607/1 Hirshfield 1961: New York USA

B. japonicum Suzuki
LB 1607/2 Suzuki 1946: type material: Japan

BODO Ehrenberg

B. caudatus (Duj.) Stein
LB 1907/4 Brooker 1963, pig dung: England

B. saltans Ehrenberg
LB 1907/2 George 1962: Wales
LB 1907/3 Brooker 1962: brackish: USA

BOEKELOVIA Nicolai *et* Baas-Becking

B. spp.
LB 908/1 Butcher 1960: marine: England
LB 908/2 Butcher 1953: marine: England
LB 908/3 Butcher 1953: marine: England
LB 908/4 Butcher 1957: marine: England
LB 908/5 Butcher: marine: England

BORODINELLA Miller

? **B. sp.**
207/1a Vischer, No. 304: *Tetracystis dissociata* Brown et Bold

BOTRYDIOPSIS Borzi

B. alpina Vischer
806/1 Vischer 1940, No. 232: soil: Switzerland

B. arrhiza Borzi
806/2 Pringsheim: soil: England
222/1b George 1947: England

B. intercedens Vischer *et* Pascher
806/3 Pringsheim: Geneva No. 522: Vischer No. 171

BOTRYDIUM Wallroth

B. becherarianum Vischer
805/1 Vischer 1937, No. 192: soil: France

B. cystosum Vischer
B 805/2 Vischer 1937, No. 196: soil: Switzerland

B. granulatum (L.) Greville
805/3a Vischer } both these strains received from Vischer in 1939
805/3b Vischer

B. granulatum *var.* **kolkwitzianum** Vischer
805/4 Vischer 1936, No. 31: from sewage: Germany

B. stoloniferum Mitra
805/5 Mitra/Pringsheim 1950: type material

BOTRYOCOCCUS Kützing

B. braunii Kütz.
LB 807/1 Droop 1950: England

BRACHIOMONAS Bohlin

B. submarina Bohlin
7/1a Droop 1950: Millport No. 42 } heterothallic pair: Finland
7/1b Droop 1950: Millport No. 43

B. submarin *var.* **pulsifera** Droop
7/2a Droop: Millport No. 44 } heterothallic pair: Scotland
7/2b Droop: Millport No. 45

BRACTEACOCCUS Tereg

B. cinnabarinus (Kol *et* Chodat) Starr
221/2 F. Chodat: Vischer No. 162: Switzerland. (formerly *Dictyococcus*)

B. engadinensis (Kol *et* Chodat) Starr
221/3 F. Chodat: Vischer No. 164: Switzerland. (formerly *Dictyococcus*)

B. minor (Chodat) Petrova
221/1 Chodat 1913 as *Botrydiopsis*, later known as *Dictyococcus*

B. minor *f.* **desertorum** Friedmann *et* Ocampo-Paus
221/6 Friedmann 1966: type material: Negev desert

B. terrestris (Kol *et* Chodat) Starr
221/4 Chodat: Vischer No. 163 (formerly *Dictyococcus*)

BULBOCHAETE Agardh

B. sp.
LB 555/1 Bold: Indiana No. 517: female strain

BUMILLERIA Borzi

B. exilis Klebs
808/2 Lewin 1951: from snow: Alaska

B. sicula Borzi
808/1 George 1950: from soil: England

B. sp.
808/3 Lewin 1951: from snow: Alaska

BUMILLERIOPSIS Printz

B. filiformis Vischer
809/2 Vischer 1943, No. 360: type material: from soil: Switzerland

B. peterseniana Vischer *et* Pascher
809/3 Vischer 1927, No. 38: type material: Switzerland

CAESPITELLA Vischer

C. pascheri Vischer = *Stigeoclonium pascheri* (Vischer) Cox *et* Bold
410/1 Vischer 1928, No. 45: type material: Switzerland
410/2 Lewin 1950: Connecticut USA

CALOTHRIX (Agardh) Bornet *et* Flahault

C. anomala Mitra *(nomen nudum)*
LB 1410/4 Mitra No. 5: type material: resembles *Scytonema* more than *Calothrix*

C. brevissima West
1410/7 Watanabe M7

C. membranacea Schmidle
1410/1 Pringsheim: material from Mitra

C. pulvinata Bornet *et* Flahault
B 1410/9 Butcher 1956: marine: England

C. scopulorum Bornet *et* Flahault
1410/5 Stewart: fixes nitrogen: Scotland

C. spp.
1410/6 Komarek/Baker 1964: mangrove: Cuba
B 1410/8 Wilcox 1971

CARTERIA Diesing

C. crucifera Korshikov
8/7a Lewin 1950: Connecticut USA
8/7b Lewin 1950: Connecticut USA
8/7c Lewin 1950: Connecticut USA

C. eugametos Mitra = *C. lunzensis* Pascher *et* Jahoda (see Fott 1968)
8/3 Mitra/Pringsheim: type material: homothallic: from soil: India

C. incisa Pringsheim *nom. prov.*
LB 8/4 Pringsheim 1941: England

C. lunzensis Pascher *et* Jahoda see *C. eugametos*

C. sp
8/5 Pringsheim 1936: Austria

CASHIA Page

C. limacoides Page

B 1534/5 Page 1965: type material: Alabama USA

CEPHALEUROS Künze

C. sp.

411/1 George 1969: isolated from leaves of tea plant showing red rust: Assam India. This is probably the causal organism though reinfection of host from the culture has not been attempted.

CERATIUM Schrank

C. hirundinella (OFM) Dujardin

LB 1110/1 Jaworski 1972, FBA 258: England

CHAETOCEROS Ehrenberg

C. calcitrans (Paulsen) Takano

LB 1010/1 Takano: a very useful form used for oyster feeding: marine: Plymouth 1056: Japan

CHAETOPELTIS Berthold

C. orbicularis Berthold

LB 412/1 Reynolds pre 1950: Wales

CHAETOPHORA Schrank

C. incrassata (Hudson) Hazen

LB 413/1 George 1949: typical material is seldom achieved in culture: England

C. sp.

LB 413/2 Reynolds No. 11

CHANTRANSIA

C. sp.

LB 1354/1 Ott, 1965, 0308 (*Audouinella*)

CHAOS Linnaeus

C. carolinense (Wilson)

LB 1511/1 Turtox/Carlsberg 13: New Jersey USA

CHARACIUM Braun

C. starrii Fott

209/1a Starr 1951: heterothallic, + strain: South Africa

209/1b Starr 1951: heterothallic, - strain: South Africa

CHILOMONAS Ehrenberg

C. curvata (auct.) Pringsheim
LB 977/1 Pringsheim 1940, strain 3: England

C. paramecium Ehrenberg
L 977/2a Pringsheim pre 1940, strain 1
LB 977/2b Hall: Pringsheim, strain 8
LB 977/2c Pringsheim, strain 4

CHLAMYDOBOTRYS Korshikov see *Pyrobotrys* Arnoldi

CHLAMYDOCAPSA Fott

C. ampla (Kütz.) Fott
31/3 Lewin 1950 (not the same as Göttingen 31/3): Connecticut USA

CHLAMYDOMONAS Ehrenberg

C. agloeformis Pascher
11/1 Mainx 1925: Czechoslovakia

C. angulosa Dill
11/59 Tsubo

C. applanata Pringsheim
11/2 Pringsheim 1930: type material: homothallic: Czechoslovakia

C. asymmetrica Korshikov
11/41 Lewin 1951: Connecticut USA

C. baca Ettl
11/77 Ettl 1960: type material: Czechoslovakia

C. brannonii Pringsheim
11/3 Brannon 1938, No. 5: USA

C. bullosa Butcher
11/83 Culture from Butcher, purporting to be Plymouth 205, but that is *Halosphaera*: reisolated by Cann 1975: marine

C. callosa Gerloff
11/24 Pringsheim 1929 (as *C. pulchra* Pringsh.): type material: Czechoslovakia

C. chlamydogama Bold
11/48a Bold 1949: + strain, W 16-1: from soil: Venezuela
11/48b Bold 1949: - strain, W 16-2: from soil: Venezuela

C. chlorostellata Flint *et* Ettl
B 11/93 Flint 1957: soil: type material: New Zealand

C. coccoides Butcher
LB 11/81 Parke 1957, Plymouth 171: possibly type material: marine: England

C. cribrum Ettl
11/75 Hindak 1962-73: type material: Czechoslovakia

C. debaryana Goroshankin

	11/40a	Lewin 1951: Connecticut USA
	11/40b	Lewin 1953: mutant No. B6: requires acetate: also mutants B1 & B9
	11/56a	Lewin 1950: soil: Mexico } heterothallic mating pair
	11/56b	Lewin 1950: soil: Mexico

C. debaryana *var.* **cristata** Ettl

	11/74	Ettl: type material: soil:Czechoslovakia

C. debaryana *f.* **Ettl**

B	11/94	Ettl 1968 as 1968/15: Czechoslovakia

C. dorsoventralis Pascher

	11/4	Mainx 1926: type material: Czechoslovakia

C. dysosmos Moewus

	11/36a	Lewin 1951: homothallic: USA
	11/36b	Lewin 1951: sterile mutant, No. 270
	11/36c	Lewin: mutant D2075: obligate autotroph
	11/31	Neish: from soil: Canada
		Also mutants D381, D2048 and D2377 from Lewin 1974

C. eugametos Moewus (see also *C. moewusii* Gerloff)

	11/5a	+ strain } isolated by Czurda from a contaminated culture from Moewus
	11/5b	- strain
	11/5c	+ strain } received from Moewus 1951
	11/5d	- strain

C. euryale Lewin

	11/62	Lewin 1957: type material: ≡ strain 11/59 of Gottingen: marine: Nova Scotia

C. fimbriata Ettl

	11/69	Hindak 1962-74: type material: Czechoslovakia

C. foveolarum Skuja

	11/68	Pringsheim 1950/Baker 1966: England

C. gerloffii Ettl

	11/72	Ettl: type material: Czechoslovakia

C. globosa Snow

	11/60	George 1954: Uganda

C. gloeopara Rodhe *et* Skuja

	11/7	Rodhe No. 1635: type material: Sweden

C. gregaria Butcher

	11/84a	Butcher/Cann 1975: marine: England
	11/84b	Butcher/Cann 1975, as 'subtype': salt marsh: Wales

C. gyrus Pascher

	11/8	Pringsheim: type material: Czechoslovakia: identity doubtful

C. humicola Lucksch

	11/9	Lucksch 1929: type material: Czechoslovakia

C. hydra Ettl

11/6a	subdioecious + strain	isolated by Czurda from contaminated cultures from Moewus
11/6b	subdioecious - strain	
11/6c	subdioecious - strain	

C. hydra Ettl *var.* micropapillata Ettl

11/76 Ettl; soil: type material: Czechlovakia

C. incisa Pringsheim (non Korsh.) see *Sphaerellopsis incisa*

C. indica Mitra

11/11 Mitra/Pringsheim 1947: soil: India

C. inepta Ettl

11/70 Ettl soil: type material: Czechoslovakia

C. intermedia Chodat

11/13 Pringsheim 1939: England

C. iyengarii Mitra

11/14 Mitra/Pringsheim 1947: soil: India

C. komma Gerloff

11/63 Tsubo as T-B-3

C. laciniata Gerloff see *Sphaerellopsis incisa*

C. mexicana Lewin

11/55a	Lewin 1953, Y62	heterothallic pair: type material: soil: Mexico
11/55b	Lewin 1953, Y64	

C. moewusii Gerloff (*C. eugametos var. moewusii* (Gerloff) Gowans) see Farooqui (1974)

11/16f	Provasoli 1948: + wild type	heterothallic pair
11/16g	Provasoli 1948: - wild type	
11/61a	Lewin 1957	heterothallic pair, not interfertile with 11/16f,g: soil: Caroline Is.
11/61b	Lewin 1957	
11/16a	Lewin, mutant M236 +: paralysed	
11/16b	Lewin, mutant M202 +: monsters	
11/16d	Lewin, mutant M470 +: twins and monsters	
11/16e	Lewin, mutant M478: slow growing, deformed plastids	

The following mutants were received from Lewin in 1974. For full reference the mutant designations should be prefixed by CCAP 11/16 in all cases.

M 224 +	M 1060 +
M 470 -	M 1074 - nm*
M 475 +	M 1081 +
M 475 nm*	M 1082+
M 1001 -	M 1083 -
M 1002 +	M 1085 -
M 1011 +	M 1086 -
M 1014 +	M 1087 -
M 1016 -	M 2002 +
M 1034 -	M 2004 +
M 1036 +	M 470.1001 +
M 1041 +	M 475.1051 +
M 1041 -	M 475.1051 -
M 1043 +	M 1001.1017 +
M 1043 -	M 1001.1074 +

* no modifier for M 1074

M 1044 +	M 1001.1074 -
M 1050 +	M 1002.1082 +
M 1051 -	M 1021.1051 +
M 1053 -	M 1021.1074 +
M 1053.1 +	M 1021.1074 -
M 1053.1 -	M 1021.1085.1 -
M 1053.2 +	M 1043.1074 -
M 1054 +	M 1043.1088 +

C. moewusii *var.* **rotunda** Tsubo
11/64a Tsubo 1952: 24 + } heterothallic pair, not interfertile
11/64b Tsubo 1952: 24 - } with any of the above: Japan

C. monoica Strehlow
11/17 Pringsheim: material from Kniep: homothallic: ? Germany

C. nivalis Wille
11/51 Lewin 1952: from snow: Alaska

C. oblonga Pringsheim
11/18 Pringsheim 1930: type material: Czechoslovakia

C. orbicularis Pringhseim
11/19 Pringsheim 1930: type material: Czechoslovakia

C. pallens Pringsheim
11/67 Pringsheim 1959: type material: South Africa

C. parkeae Ettl
11/79 Parke/Cann: type material: marine: Plymouth 285: England

C. perforata Pascher *et* Jahoda
11/43 Pringsheim 1950: Finland

C. philotes Lewin
11/53 Lewin 1953: type material: from soil: homothallic: Mexico

C. plethora Butcher
11/84a Butcher/Cann 1975: marine: England
11/86a Butcher/Cann 1975: designated "sub-type" by Butcher: brackish: England
11/86b Butcher/Cann 1975: brackish: England

C. proboscigera *var.* **conferta** Ettl
11/38 Lewin 1951: Connecticut USA

C. proteus Pringsheim
11/21 Pringsheim 1930: type material: Czechoslovakia

C. pseudagloe Pascher
11/22b Lewin 1950: Connecticut USA

C. pseudococcum Lucksch
11/23 Lucksch 1929: type material: Czechoslovakia

C. pulchra Pringsheim (= *C. callosa* Gerloff)

C. pulsatilla Wollenweber
11/44 Pringsheim 1950: Finland

C. pulvinata Vischer

11/25	Vischer 1923, No. 4: type material: Switzerland

C. rapa Ettl *f.* **vasta** Ettl

11/73	Ettl 1962: type material: Czechoslovakia

C. reginae Ettl *et* Green

11/78	Jowett 1966/Cann 1975: Plymouth No. 399; type material: France

C. reinhardii Dangeard

11/32a	G. M. Smith No. 137c + } from soil: Massachusetts USA
11/32b	G. M. Smith No. 137c - } from soil: Massachusetts USA
11/32c	wild type + { from R. P. Levine: morphologically identical
11/32d	wild type - { with preceding pair but not interfertile
11/32e	Levine, mutant Arg. 1 -
11/32f	Levine, mutant Arg. 2 +
11/32g	Levine, mutant Pab. 2 +
11/32h	Levine, mutant Pf. 15A +
11/32k	Levine, mutant Pab. 1A +
11/32m	Levine, mutant Ac. 17 +
11/32n	Levine, mutant Thi. 2 -
11/32o	Levine, mutant Thi. 8 -
11/32p	Levine, mutant Pf 1 +
11/32q	Levine, mutant Ac. 31 +
11/32r	Levine, mutant Pf. 14 -
11/32s	Levine, mutant SR. 1a +
11/32w	Levine, mutant Ac. 14 A Arg. 2 -, allelic
11/32x	Levine, mutant Ac. 14 B +, allelic
11/32y	Levine, mutant Ac. 14 C -, allelic
11/32z	Levine, mutant Ac. 14 E -, allelic
11/45	Lewin 1950: (*C. reinhardii?*): Connecticut USA

Owing to the large number of mutant strains involved, we are using the isolator's designations. The full reference to these strains should, therefore, be CCAP 11/32 Pf 1-, etc.

Pf 1 -	R. A. Lewin	HA 24 -	McVittie
Pf 2 +	R. A. Lewin	HA 26 -	McVittie
Pf 3 +	R. A. Lewin	HA 28 -	McVittie
Pf 4 +	R. A. Lewin	HA 29 +	McVittie
Pf 5 -	R. A. Lewin	NG 1 +	McVittie
Pf 6 -	R. A. Lewin	NG 2 +	McVittie
Pf 7 -	R. A. Lewin	NG 5 + (Lf1)	McVittie
Pf 8 +	R. A. Lewin	NG 6 +	McVittie
Pf 9 -	R. A. Lewin	NG 9 +	McVittie
Pf 10 +	R. A. Lewin	NG 10 +	McVittie
Pf 12	R. A. Lewin	NG 11 -	McVittie
Pf 13 -	R. A. Lewin	NG 13 + (Lf2)	McVittie
Pf 14 +	R. A. Lewin	NG 14 +	McVittie
Pf 16 -	R. A. Lewin	NG 15 +	McVittie
Pf 17 -	R. A. Lewin	NG 16 +	McVittie
Pf 18 +	R. A. Lewin	NG 17 +	McVittie
Pf 19 +	R. A. Lewin	NG 19 +	McVittie
Pf 20 +	R. A. Lewin	NG 22 +	McVittie
AO 3 +	McVittie	NG 24 +	McVittie
AO 4 +	McVittie	NG 25 +	McVittie
AO 6 +	McVittie	NG 28 +	McVittie
AO 7 +	McVittie	NG 30 +	McVittie

HA 1 - McVittie
HA 3 - McVittie
HA 7 McVittie
HA 10 - McVittie
HA 13 - McVittie
HA 16 - McVittie
HA 18 + McVittie
HA 19 + McVittie
HA 21 - McVittie
NG 35 + McVittie
NG 36 + McVittie
NG 37 + McVittie
S 3 + McVittie
S 13 + McVittie
S 16 + McVittie
SU 1 J. R. Warr

Mutants received from D. R. Davies 1974/75

Ac 14 E-
Ac 14 ECW1+
Ac 14 ECW2
Ac 14 ECW8
Ac 14 ECW10
Ac 14 ECW18/2+
Ac 14 ECW 18/3-
Ac 14 ECW18/4+
Ac 14 ECW18/6-
Ac 14 ECW19
Ac 14 E UVS1-
Ac 14 E UVS1/B-
2 Ac 14 EY/1
Ac 17 E-
Ac 17 ECW8
Ac 17 ECW10
Ac 76+
Arg 1+
Arg 2+
Can R+
CW 1-
CW 2+
CW 3+
CW 4+
CW 6+
CW 7-
CW 8+
CW 9-
CW 10-
CW 14+
CW 15+
CW 17-
CW 18+
CW 18-
CW 19-
CW 20-
CW 51-
CW 51+
CW 92-
CW 177+
Nic 13+
Pab 1+
Pab 2-
PF 4-
SR 1+
UVS 1-
UVS1/b+
UVS 2+
UVS 3-
UVS 4-
UVS 5+
UVS 6+
UVS 7-
UVS 8
UVS 9
Y +
YUVS 1+
YUVS 6+

C. rosae H. *et* O. Ettl (= *Chloromonas rosae* Ettl)
11/66 Ettl: type material: from soil: Czechoslovakia

C. ? rotula Playfair
11/33 Vischer, No. 389

C. rugosa Butcher
11/85 Butcher/Cann 1975: marine

C. segnis Ettl
11/71 Ettl: soil: type material: Czechoslovakia

C. aff. snowiae Prinz
8/2 arose as a contaminant in a *Carteria* culture

C. sphaerella Pringsheim
11/27 Pringsheim: England

C. sphaeroides Gerloff
11/29 Pringsheim 1929: Czechoslovakia

C. stercoraria Pringsheim *nom. prov.*
11/49 Pringsheim 1951: England

C. subangulosa Fritsch *et* John
11/28 Pringsheim 1940: material from John: from soil: England

C. spreta Butcher
LB 11/87 Butcher: marine: England

C. subehrenbergii Butcher
11/88 Butcher/Cann 1975: marine: England

C. subglobosa Pringsheim (= *C. sphaeroides* Gerloff)
11/29 Pringsheim 1929: type material: Czechoslovakia

C. subtilis Pringsheim
11/30 Pringsheim 1929: type material: Czechoslovakia

C. terricola Gerloff
11/37 Lewin 1950: Connecticut USA

C. ulvaensis Lewin
11/58 Lewin 1951: type material: Scotland

C. uva-maris Butcher
11/89 Butcher/Cann 1975: marine: England

C. vectensis Butcher
11/90 Butcher/Cann 1975: marine: England

C. spp.

	11/26	Pringsheim: formerly listed as *C. simplex*
	11/35	Lewin 1950: marine: used for oyster feeding: USA
	11/42	Lewin 1951: *(Chlorogoniella)*: Massachusetts USA
LB	11/46	George 1948: *(Aglöe)*: England
	11/47	Droop 1951: Finland
	11/50	Droop 1951: *(Chloromonas)*: Finland
	11/52	Lewin 1952: from snow: Alaska USA
	11/54	Lewin 1953, Y 80: homothallic: from soil: Mexico
	11/56a	Lewin 1953, Y 68 } mating pair : from soil: Mexico
	11/56b	Lewin 1953, Y 69 } mating pair : from soil: Mexico
	11/57	Lewin 1951, DD 1/174: homothallic: Scotland
	11/82	Blakey 1973: *(Euchlamydomonas)*: a large form recommended for teaching: England
	11/91	Butcher/Cann 1975: marine: England

CHLORELLA Beijerinck

It has long been recognised that the specific names given to strains of *Chlorella* have in many cases been useless or even misleading. The names *C. pyrenoidosa* and *C. vulgaris* especially have been used with little regard for either the characters of the alga or the code and practice of nomenclature. In general, we follow the revision of Fott and Novakova (1969) as modified by Kessler (1974) and confirmed by Morris (1976). Unfortunately, over the years, there has been some confusion of strains in all the main culture collections. Where the Göttingen strains have proved different from ours of the same designation, we have distinguished them by adding 'Gö' or 'CCAP' to the strain number.

All except the marine strains are now preserved under liquid nitrogen, indicated by 'N'. Only the strains in frequent demand will still be maintained on agar slopes. The marine species have not yet been taxonomically revised.

Chlorella autotrophica Shihira *et* Krauss
=*C. vulgaris var. autotrophica* (S. *et* K.) Fott *et* Novakova

C. candida S. *et* K.
=*C. vulgaris var. vulgaris* according to F. *et* N.

C. ellipsoidea Gerneck
=*C. saccharophila var. ellipsoidea* (Gerneck) F. *et* N.

C. emersonii S. *et* K. *var.* **emersonii** (formerly *C. fusca var. vacuolata.* see Fott *et al.*, 1975)

N	211/8a	Pringsheim: type material of *C. photophila* S. *et* K.
N	211/8b	Emerson 1923: type material of *C. fusca var. vacuolata*, formerly known as *C. pyrenoidosa*: Pa USA
N	211/8c	Emerson 1926: from tapwater in Warburg's laboratory: formerly known as *C. pyrenoidosa*: Germany
N	211/8g	Emerson '3': used in Cambridge Botany School as *C. pyrenoidosa* up to 1957
N	211/8h	Emerson '3': received from A. H. Brown 1957
N	211/11m	'Cornell 11' used by *inter alia* Hopkins and Wann (1926) and Pearsall and Loose (1937): from soil
N	211/11n	from Emerson's laboratory, isolated before 1939
N	211/15	Pringsheim c. 1945: England

C. emersonii *var.* **rubescens** Fott *et al.*

N	232/1	Izard 1966: type material of *Halochlorella rubescens* Dangeard

C. fusca S. *et* K. *var. fusca* = *Scenedesmus spp.* see Fott, B., Lochhead *et* Clemencon (1975)

C. fusca var. rubescens (Dangeard) Kessler *et al* = *C. emersonii var. rubescens* Fott *et al*

C. fusca var. vacuolata S. *et* K. = *C. emersonii* S. *et* K. *var. emersonii*

C. kessleri F. *et* N.

N	211/11g	Winokur, received 1945: type material: also type material of *C. regularis*
N	211/11h	used by Pratt, received from Emerson 1946.

C. luteoviridis Chodat

N	211/2a	Kufferath 1912: Geneva No. 95: type material: Belgium
N	211/2b	Beijerinck, received from Baarn 1949
N	211/3	? Kluyver: from Delft as *C. aureoviridis*
N	211/4	Kufferath 1912: Geneva No. 111: type material of *var. lutescens* Chodat, Belgium
N	211/5a	Pringsheim: type material of *C. mutabilis* S. *et* K.
N	211/5b	Gaffron: HMS No. Z.11.1.1 as *Palmellococcus*: *C. nocturna* S. *et* K.
N	211/10a (CCAP)	Beijerinck: Meyer No. 20, clone B 410
N	211/10d (CCAP)	Beijerinck: received from Delft 1946
N	211/10e (CCAP)	received from Prague 1946

C. marina Butcher

LB	211/27	Collyer: marine: type material?: England

C. miniata (Naegeli) Oltmanns

N	211/14	Dönz 1933: type material: type of *C. zofingiensis* according to F.*et* N: Switzerland

C. mutabilis S. *et* K. = *C. luteoviridis* Chodat

C. nocturna S. *et* K. See *C. luteoviridis*

C. ovalis Butcher = *C. vulgaris var. vulgaris*

N	211/21a	Butcher/George 1956: type material: marine: England
LB	211/21b	Butcher 1953: marine: England

C. photophila S. *et* K.

	211/8a	Pringsheim: type material: = *C. emersonii var. emersonii*

C. pringsheimii S. *et* K. = *C. saccharophila*

C. protothecoides Krüger

N	211/7a	Krüger 1912: type material
N	211/7b	? Kluyver
N	211/7c	? Kluyver
N	211/7d	Czurda: type material of *var. galactophila* S. *et* K.
N	211/8d	Pringsheim 1947: from *Ulmus* sap: England
N	211/10a (Gö)	
N	211/10b	Beijerinck: Chodat No. 21, Clone G 13
N	211/10c	Beijerinck: Chodat No. 21, Clone G 122
N	211/10d (Gö)	
N	211/10e (Gö)	
N	211/11a	Pringsheim: *var. mannophila* S. *et* K.
N	211/11i	Beijerinck, received from Baarn 1946: selected as type of *C. vulgaris* Beij. by Drouet *et* Daily (1956)
N	211/13	Beijerinck, as *C. xanthella*: received from Delft 1948
N	211/17	Beijerinck, from Baarn 1949
N	211/6	Loefer 1934: from *Paramecium bursaria*

C. pyrenoidosa Chick *nomen confusum*. Several strains used under this name are now listed as *C. emersonii*

C. regularis (Artari) Oltmanns *nom. illeg.*: = *C. kessleri*

C. saccharophila (Krüger) Nadson

N	211/1a	Pringsheim: type material of *var. ellipsoidea* (Gerneck) F. *et* N.
N	211/1b	Brannon 1938, No. 6: USA
N	211/1c (Gö)	
N	211/1d	Kellner 1951: from Pirson as M-St
N	211/1f	Wood, Millport 116
N	211/9a (Gö)	
N	211/9b (Gö)	

C. salina Kufferath

LB	211/25	Parke 1963, Plymouth No. 309: marine: England

C. sorokiniana S. *et* K. = *C. vulgaris f. tertia* F. *et* N.

C. spaerckii Ålvik

LB	211/29a	Butcher: marine: Wales
LB	211/29b	Butcher: marine: Wales

C. stigmatophora Butcher

	211/20	Parke 1935/George 1954: type material: marine: Plymouth No. 85, off Isle of Man

C. variabilis S. *et* K. = *C. protothecoides*

C. variegata Beijerinck *nom. nud.* = *C. protothecoides*

C. vulgaris Beijerinck *var.* **vulgaris**

N	211/1c (CCAP)	Rodhe No. 1633
N	211/1e	Gaffron: used by Tamiya as A002
N	211/9a (CCAP)	Kruger 1892: type material of *Chlorothecium saccharophilum*
N	211/9b (CCAP)	Pringsheim, ? 1939: England
N	211/11b	Beijerinck: type material: = *C. candida* according to S. *et* K.
N	211/11j	Rodhe No. 1630: type material of *C. simplex* S. *et* K.
N	211/11p	Algéus 1942: 'B' Sweden
N	211/11q	Czurda
N	211/11s	from Pirson 1952 as M.A.
N	211/12	Chodat: type material of *var. viridis* Chodat, Geneva No. 45
N	211/19	received from Harder 1950
N	211/11c	Pringsheim: type material of *C. candida* S. *et* K.
N	211/11f	Pringsheim: 1939: England
N	211/11r	Krollpfeiffer: received from Pirson 1952 as M-Kr
N	211/21a	Butcher/George 1956: type material of *C. ovalis*: marine: England

C. vulgaris *var.* **autotrophica** (S. *et* K.) F. *et* N

N	211/24	Lewin: University of Indiana No. 580

C. vulgaris *f.* **tertia** F. *et* N

N	211/8k	Sorokin *et* Myers, Tx 7.11.05: high temperature strain: (*C. sorokiniana* S. *et* K.) Texas USA
N	211/11d	Brannon 1938, No. 1: USA
N	211/11k	Rice
N	211/18	Algéus 1948, 'A': Sweden

C. vulgaris var. viridis Chodat = *C. vulgaris var. vulgaris* according to F. *et* N.

C. xanthella Beijerinck *nom. nud.* = *C. protothecoides*

C. zofingiensis Dönz

N	211/14	Donz 1933: type material: also type of *C. miniata* S.*et* K: Switzerland

C. spp.

N	211/8p	derived from Cornell 11, now differs slightly: received from Syrett 1965
N	211/9c	Thain 1970, Endophyte from *Selaginella*: Australia
N	211/22	Lewin 1957, Endozoon from *Spongilla fluviatilis*: USA
N	211/26	Fogg 1956: arctic: Sweden
	211/28	Ho 1972, UL 196 RI 200674: factory effluent, Malaya

CHLORELLA - Numerical list

211/1a		*C. saccharophila*
211/1b		*C. saccharophila*
211/1c		*C. vulgaris*
211/1c	Gö	*C. saccharophila*
211/1d		*C. saccharophila*
211/1e		*C. vulgaris*
211/1f		*C. saccharophila*
211/2a		*C. luteoviridis*
211/2b		*C. luteoviridis*
211/3		*C. luteoviridis*
211/4		*C. luteoviridis*
211/5a		*C. luteoviridis*
211/5b		*C. luteoviridis*
211/6		*C. protothecoides*
211/7a		*C. protothecoides*
211/7b		*C. protothecoides*
211/7c		*C. protothecoides*
211/7d		*C. protothecoides*
211/8a		*C. emersonii*
211/8b		*C. emersonii*
211/8c		*C. emersonii*
211/8d		*C. protothecoides*
211/8g		*C. emersonii*
211/8h		*C. emersonii*
211/8k		*C. vulgaris f. tertia*
211/8p		*C. sp.*
211/9a		*C. vulgaris*
211/9a	Gö	*C. saccharophila*
211/9b		*C. vulgaris*
211/9b	Gö	*C. saccharophila*
211/9c		*C. sp.*
211/10a		*C. luteoviridis*
211/10a	Gö	*C. protothecoides*
211/10b		*C. protothecoides*
211/10c		*C. protothecoides*
211/10d		*C. luteoviridis*
211/10d	Gö	*C. protothecoides*
211/10e		*C. luteoviridis*
211/10e	Gö	*C. protothecoides*
211/11a		*C. protothecoides*
211/11b		*C. vulgaris*
211/11c		*C. vulgaris*
211/11d		*C. vulgaris f. tertia*
211/11f		*C. vulgaris*
211/11g		*C. kessleri*
211/11h		*C. kessleri*
211/11i		*C. protothecoides*
211/11j		*C. vulgaris*
211/11k		*C. vulgaris f. tertia*
211/11m		*C. emersonii*
211/11n		*C. emersonii*
211/11p		*C. vulgaris*
211/11q		*C. vulgaris*
211/11r		*C. vulgaris*
211/11s		*C. vulgaris*
211/12		*C. vulgaris*
211/13		*C. protothecoides*
211/14		*C. zofingiensis*
211/15		*C. emersonii*
211/17		*C. protothecoides*
211/18		*C. vulgaris f. tertia*
211/19		*C. vulgaris*
211/20		*C. stigmatophora*
211/21		*C. vulgaris*
211/22		*C. spp.*
211/23		*Scenedesmus sp.*
211/24		*C. vulgaris var. autotrophica*
211/25		*C. salina*
211/26		*C. sp.*
211/27		*C. marina*
211/28		*C. sp.*
211/29a		*C. spaerckii*
211/29b		*C. spaerckii*

CHLORELLIDIUM Vischer *et* Pascher

C. tetrabotrys Vischer *et* Pascher

811/1a Vischer No. 153: Switzerland
811/1b Vischer No. 155: Czechoslovakia

CHLORHORMIDIUM Fott = *Klebsormidium* Silva, Mattox *et* Blackwell

CHLORIDELLA Pascher

C. minuta Moewus

813/2 Moewus: type material

C. neglecta Pascher

813/1 Vischer 1940, No. 216: from soil: Switzerland

CHLOROBOTRYS Bohlin

C. regularis (W. West) Bohlin

LB 810/1 Hibberd 1967: England

CHLOROCHYTRIDION Vischer

C. tuberculata Vischer see *Pedinomonas tuberculata*

CHLOROCHYTRIUM Cohn

C. spp.

212/1 Mitra/Pringsheim
LB 212/2 George 1952: England

CHLOROCLOSTER Pascher

C. engadinensis Vischer

812/1 Vischer 1940, No. 252: from soil: Switzerland

C. solani George see *Pseudococcomyxa adhaerens*

Our strains of *Chlorocloster* were shown to contain chlorophyll b. They are presumably not therefore members of the Xanthophyceae. Until the type species is checked, *Chlorocloster* may itself still be a xanthophycean genus.

CHLOROCOCCUM Meneghini

C. echinozygotum Starr

213/5 Bold: type material: from soil: Philippine Is.

C. hypnosporum Starr

213/6 Bold: type material: from soil: Tennessee USA
237/1 Pringsheim 1940: formerly listed as *Hypnomonas lobata*
Korsh: soil: England

C. infusionum (Schrank) Meneghini

	213/2a	Mainx
	213/2b	Mainx

C. macrostigmatum Starr

	213/9	Starr: from soil: Connecticut USA

C. minutum Starr

	213/7	Bold: type material: from soil: India

C. multinucleatum Starr = *Neospongiococcum granatum* Deason

	213/1a	Pringsheim (as *C. humicolum*): type material: from soil

C. vacuolatum Starr

	213/8	Starr 1952: type material: from soil: South Africa

C. wimmeri Rabenhorst = *Neochloris wimmeri* (Rab.) Archibald *et* Bold

	213/4	Mainx: Meyer No. Chlor. 7

CHLOROGLOEA Wille

C. fritschii Mitra (= ? *Nostoc sp.*)

LB	1411/1a	Mitra 1950: No. 7: type material: from soil: India

C. sp.

L	1411/2	Kunisawa, Stanier 6712: California USA

CHLOROGLOEOPSIS Mitra *et* Pandy

C. sp. *"Chlorogloea fritschii"*

	1411/1b	origin previous to Botany Department, Berkeley, is uncertain: Stanier 6718: fixes nitrogen

CHLOROGONIUM Ehrenberg

C. elongatum Dangeard

	12/1	Pringsheim 1942: England
	12/2a	Hartmann/Pringsheim
	12/2b	George 1948: South Africa
	12/2c	Pringsheim 1949: France
	12/2d	Pringsheim 1951: South Africa
	12/4	Meyer: formerly listed as *C. tetragamum*: Czechoslovakia

C. euchlorum Ehrenberg

	12/3	Kniep/Mainx: Germany

C. spp.

LB	12/5	Pringsheim as K
L	12/6	Pringsheim as N

CHLOROMESON Pascher

C. sp.

LB	814/1	Butcher 1954: marine: England

CHLOROSARCINOPSIS Herndon

C. negevensis Friedmann *et* Ocampo-Paus *f.* **ferruginea**
14/1 Baldinger 1-112: type material: Negev Desert

C. negevensis *f.* **negevensis**
14/2 Baldinger 1-116: type material: Negev Desert

C. sempervirens Groover *et* Bold
214/1 Pringsheim: type material, probably isolated in Prague before 1939

CHLOROSPHAERA Klebs

C. klebsii Vischer
215/1 Vischer, No. 58

CHROMULINA Cienkowski

C. ochromonoides (Auct?)
LB 909/1 Butcher 1959: marine: England
LB 909/3 Butcher 1956: marine: England

C. spp.
LB 909/2 Butcher 1956: marine: England
LB 909/5 Butcher: marine: Wales
LB 909/6 Butcher 1957: marine: England
LB 909/7 Butcher 1956: marine: England
LB 909/8 Butcher: marine: England

CHROOCOCCOPSIS Geitler

C. sp.
B 1409/1 Butcher: marine: England

CHROOCOCCUS Naegeli

C. minutus (Kütz.) Naeg.
B 1412/5 Laporte 1965: thermal water

C. prescottii Drouet *et* Daily
LB 1412/4 George 1954: Hong Kong

C. turgidus (Kütz.) Naegeli
LB 1412/1b Starr (a large form)

C. ? versicolor
LB 1412/2 Pringsheim (a very small form: ?*Coccochloris peniocystis*)

CHROOMONAS Hansgirg

C. atrorosea Butcher
LB 978/6a Butcher: type material: marine: England
LB 978/6b Butcher: marine: England

C. collegionis Butcher
LB 978/11 Butcher 1961: type material: marine: England

C. diplococca Butcher
LB 978/26 Butcher 1960: marine: England

C. dispersa Butcher
LB 978/10 Butcher 1960: type material: marine: England

C. falcata Butcher
LB 978/5a Butcher: type material: marine: Wales
LB 978/5b Butcher: marine: England

C. heteromorpha Butcher
LB 978/7 Butcher 1960: marine: England

C. mesostigmatica Butcher
LB 978/9 Butcher: type material: marine: England

C. placoidea Butcher
LB 978/8 Butcher 1959: type material: marine: England

C. salina (Wislouch) Butcher
LB 978/24 Butcher 1959: marine: England
LB 978/27 Butcher 1954: marine: England

C. salina *f.* adolescens Butcher
LB 978/12a Butcher 1957: marine: England
LB 978/12b Butcher 1958: marine: England

C. salina *f.* carterae Butcher
LB 978/13 Butcher 1956: marine: England
LB 978/14 Butcher 1958: marine: England

C. salina *f.* eroticon Butcher
LB 978/15a Butcher: marine: England
LB 978/15b Butcher: marine: England

C. salina *f.* granulata Butcher
LB 978/16 Butcher 1954: marine: England
LB 978/17 Butcher: marine: England

C. salina *f.* leucofera Butcher
LB 978/19 Butcher 1955: marine: England
LB 978/20 Butcher: marine: Wales

C. salina *f.* oculus-bovis Butcher
LB 978/21a Butcher 1957: marine: England
LB 978/21b Butcher 1959: marine: England

C. salina *f.* refracta Butcher
LB 978/22 Butcher 1956: marine: England
LB 978/23 Butcher: marine: England

C. virescens (Butcher) Butcher
LB 978/25 Butcher 1958: marine: England

C. spp.
LB 978/1 Pringsheim, 1
LB 978/2 Pringsheim, 2
LB 978/3 Pringsheim, 'Wales'
LB 978/4 Pringsheim, 4

CHROOTHECE Hansgirg

C. richterianum Hansgirg (= ? *Asterocytis ornata* (Ag.) Hamel see Drouet and Daily 1956, p147)
B 1353/1 Belcher 1956: brackish? : Isle of Man

CHRYSOCAMPANULA Fournier

C. spinifera Fournier
LB 907/1 Jowett 1964, Plymouth No. 328: marine: England

CHRYSOCHROMULINA Lackey

C. acantha Leadbeater *et* Manton
LB 910/10 Parke 1964, Plymouth No. 326: type material: marine: England

C. chiton Parke *et* Manton
LB 910/7 Parke 1955, Plymouth No. 146: type material: marine : England

C. cymbium Leadbeater *et* Manton
LB 910/13 Parke 1964: type material: marine: England

C. ericina Parke *et* Manton
LB 910/4 Parke 1950, Plymouth No. 25: type material: marine : England

C. kappa Parke *et* Manton
LB 910/1 Parke 1939, Plymouth No. K: type material: marine : Isle of Man

C. minor Parke *et* Manton
LB 910/3 Parke, Plymouth No. 52: type material: marine

C. polylepis Manton *et* Parke
LB 910/9 Adams 1958, Plymouth No. 200: marine: England

C. pringsheimii Parke *et* Manton
LB 910/11 Parke 1957, Plymouth No. 166: marine: England

C. strobilus Parke *et* Manton
LB 910/12 Jowett 1967, Plymouth No. 43a: marine: England

CHRYSOCOCCUS Klebs

C. rufescens Klebs
LB 916/1 Belcher and Swale 1976: England

CHRYSOSPHAERA Pascher

C. magna Belcher
LB 911/1 Belcher 1972: type material: England

CLADOPHORA Kützing

C. coelothrix Kützing
LB 505/10 van den Hoek 1960: marine: Algeria

C. fracta Kützing *var.* **fracta**
LB 505/1b George 1950: Sweden
LB 505/2a George 1948: England
LB 505/2b van den Hoek 1960: Holland

C. fracta *var.* **intricata** (Lyngbye) van den Hoek
LB 505/1a George 1947: England

C. globulina Kützing
LB 505/5 van den Hoek 1961: France

C. glomerata (L.) Kützing
LB 505/3 George 1950: rarely typical in culture: England

C. glomerata *var.* **crassior** (Ag.) van den Hoek
LB 505/4 van den Hoek 1959: Holland

C. kosterae van den Hoek
LB 505/6 van den Hoek 1961: type material: France

C. parriaudii van den Hoek
LB 505/9 van den Hoek 1960: type material: marine:France

C. sp.
LB 505/11 Westlake 1957: England

CLOSTERIUM Ralfs

C. acerosum (Schrank) Ehrenberg
LB 611/4 George 1951: England

C. braunii Reinsch
LB 611/7 Lefèvre 1954: France

C. ehrenbergii Menegh. *var.* **malinvernianum** De Not
LB 611/8 Hibberd 1970: used for demonstration

C. leibleinii Kützing
LB 611/2 Starr 1951: France

C. littorale Gay
LB 611/6 Starr 1956: homothallic: Indiana No. 736

C. moniliferum Ehrenberg
LB 611/1 Starr 1951: France

C. parvulum Naegeli
LB 611/5 Starr: homothallic: Indiana No. 735

COCCOCHLORIS Sprengel

C. elabens (Bréb.) Drouet *et* Daily
LB 1413/1 Lewin: saline: California USA

C. peniocystis (Kütz.) Drouet *et* Daily
LB 1412/2 Pringsheim as ? *Chroococcus versicolor*
LB 1461/1 Pringsheim 1948 as *Pelogloea*: England

COCCOLITHUS Schwarz

C. pelagicus (Wallich) Schiller
LB 913/2 Parke, Plymouth No. 182: marine: English Channel
LB 913/3 Parke 1964, Plymouth No. 317: marine: English Channel
LB 912/1 Pringsheim: marine: Isle of Man

C. spp.
LB 913/4a Butcher 1961: marine: England
LB 913/4b Butcher 1960: marine: England
LB 913/5 Butcher 1956: marine: with type material of *Paraphysomonas butcheri*: England
LB 913/6 Butcher 1961: marine: Guernsey
LB 913/7 Butcher: marine: with colourless flagellate: England

COCCOMYXA Schmidle

C. arvernensis Jaag
216/1 Jaag, No. 149: type material: France

C. chodatii Jaag
216/2 Chodat, No. 186: type material: Switzerland

C. elongata Jaag
216/3a Pringsheim 1927: identified by Jaag, No. 203: Czechoslovakia
216/3b Chodat, No. 190: Switzerland
216/3c Brannon 1938, No. 3: USA

C. mucigena Jaag
216/4 Jaag, No. 123: *Peltigerae aphtosae*: Switzerland

C. peltigerae Waren
216/5 Jaag

C. peltigerae variolosae Jaag
216/6 Jaag, No. 194: type material: Norway

C. pringsheimii botrydinae Jaag
216/7 Pringsheim: Jaag, No. 202: type material: Finland

C. rayssiae Chodat *et* Jaag
216/8 Rayss: Jaag, No. 185: Rumania

C. simplex (Pringsheim) Mainx
216/9a Pringsheim: Czechoslovakia
216/9b Mainx: Germany

C. solorinae bisporae Jaag

216/10 Jaag, No. 142: type material: Switzerland

C. solorinae croceae Chodat

216/11a Chodat, Jaag, No. 188: type material: Switzerland
216/11b Jaag, No. 143: Switzerland

C. solorinae saccatae Chodat

216/12 Jaag, No. 189: Switzerland

C. subellipsoidea Acton

216/13 Pringsheim (*botrydinae*): lichen originally from Skuja
216/15 Richardson 1964 *ex Botrydina vulgaris*

C. viridis Chodat

216/14 Chodat: Jaag, no. 192: type material: Switzerland

C. spp.

216/16 Bednar *ex Peltigera aphtosa*: USA
LB 216/17 Butcher: marine

COCHLIOPODIUM Hertwig *et* Lesser

C. actinophorum (Auerbach)

B 1537/2 Page 1964: Wisconsin USA

C. actinophorum *var.* **minus** Page

B 1537/1a Page 1965: type material: Wisconsin USA

COELASTRELLA Chodat

C. costata (Korshikov) Kalina

LB 217/5 Droop 1950, as *Coelastrum morus*: Finland

COELASTRUM Naegeli

C. microporum Naegeli

217/1a Pringsheim 1940: England
217/1b Rodhe, No. 1615a; Sweden
217/1c Starr 1951: England

C. morus W. *et* G. S. West

217/4 Lewin 1950? No. 6: Connecticut USA

C. proboscideum Bohlin *var.* **dilatatum** Vischer

217/2 Vischer 1924, No. 13: type material: Switzerland

C. proboscideum *var.* **gracile** Vischer

217/3 Vischer 1924, No. 15: type material: Switzerland

COELOSPHAERIUM Naegeli

C. kutzingianum Naegeli

LB 1414/1 George 1951: England

COENOCOCCUS Korshikov

C. planctonicus Korshikov
LB 112/1 Lund 1967, FBA L126: England

COLACIUM Ehrenberg

C. cyclopicolum (Gicklhorn) Pringsheim (= *C. vesiculosum* Ehr. *q.v.*)

C. mucronatum Bourrelly *et* Chadefaud
1211/2 Bourrelly, Paris, No. 100

C. vesiculosum Ehrenberg
1211/2 Pringsheim 1943: off *Daphnia*: England

COLEOCHAETE Brébisson

C. scutata Brébisson
LB 414/1 George 1949: England

COLEPS Nitzsch

C. hirtus OFM
LB 1613/1 George 1950 (cultured with *Chilomonas*): England
LB 1613/2 George 1965 (cultured with *Chilomonas*): England

COLPIDIUM Stein

C. colpidium (Schewiakoff) Corliss
LB 1614/1 George 1953: Ireland

C. striatum Stokes
LB 1614/2 Jankowsky: Russia

COLPODA Müller

C. aspera Kahl
LB 1615/3 Page 1974: soil: England

C. cucullus Müller
LB 1615/2 West 1971: England

COSMARIUM Ralfs

C. bioculatum Brébisson
LB 612/17 Jaworski 1967, FBA L129: England

C. botrytis Meneghini
612/1a Pringsheim/Czurda 1926: Germany
612/4 Ondraček, No. 1: Prague No. 227
612/5 Ondraček, No. 2: Prague No. 228

C. botrytis *var.* **depressum** W. *et* G. S. West
LB 612/1b Christensen 1956, No. 5614: homothallic: Indiana No. 953

C. **contractum** Kirchner *var.* **ellipsoideum** (Elfv.) West *et* West

LB	612/16	Lund 1958, FBA L128: England

C. **cucumis** Corda

	612/10	Ondraček, No. 27: Prague No. 239

C. **formulosum** Hoff

	612/7	Ondraček, No. 8: Prague No. 232 (a small form)

C. **impressulum** Elfv.

	612/2	Czurda 1925: Czechoslovakia

C. **lundellii** Delponte

LB	612/15	Lefèvre 1929: France

C. **praemorsum** Brébisson

B	612/13	George 1951: England

C. **subtumidum** Nordsted

	612/8a	Ondraček No. 25: Prague No. 245
	612/8b	Ondraček No. 26: Prague No. 246
	612/11	Ondraček No. 22: Prague No. 242, differs slightly from type
	612/12	Lewin 1950: differs slightly from type: Connecticut USA

C. **turpinii** Brébisson

LB	612/14a	Starr No. 3: Indiana No. 733 } heterothallic pair
LB	612/14b	Starr No. 4: Indiana No. 734

CRICOSPHAERA Braarud

C. **carterae** (Braarud *et* Fagerland) Braarud (formerly *Syracosphaera)*

LB	961/1	Parke 1949, Plymouth No. 17: marine: England
	944/1	Pringsheim 1951: type material of *Pleurochrysis scherffelii*: brackish: England

C. *aff.* **carterae**

LB	961/2	Adams 1958, Plymouth No. 181: marine: England
LB	961/5	Adams 1956, Plymouth No. 156: marine: a good food organism: England

C. **elongata** (Droop) Braarud

LB	961/3	Droop, Millport 62, Plymouth No. 414: type material: marine: England

C. **sp.**

LB	961/4	von Stosch, No. 7133 as *Syracosphaera sp* : Plymouth No. 151: marine

CRUCIGENIA Morren

C. **tetrapedia** (Kirchner) W. *et* G. S. West

	218/3	Starr: Indiana No. 63: Massachusetts USA

CRUCIGENIELLA Lemmermann

C. **apiculata** (Lemmermann) Komárek

	218/1	George 1949: from Paris No. 89

C. **rectangularis** (Naeg.) Komárek

	218/2	Pringsheim 1951: France

CRYPTODIFFLUGIA Penard

C. **oviformis** Penard

B	1514/1	Page 1964: type material of *C. operculata* Page: Wisconsin USA
B	1514/2	Hedley 1968: Wales

CRYPTOGLENA Ehrenberg

C. **pigra** Ehrenberg

LB	1212/1	Pringsheim 1947: England

CRYPTOMONAS Ehrenberg

C. **abbreviata** Butcher

LB	979/16	Butcher 1960: marine: Guernsey

C. **acuta** Butcher

LB	979/10	Butcher: type material: marine: Wales

C. **appendiculata** Butcher

LB	979/13	Parke 1950, Plymouth No. 23: marine: Scotland

C. **calceiformis** Lucas

LB	979/6	Jowett 1966, Plymouth No. 412: type material: marine: England

C. **chrysoidea** Butcher

LB	979/8	Butcher 1953: type material: brackish: England

C. **irregularis** Butcher

LB	979/7	type material: marine: England

C. **maculata** Butcher

LB	979/12	Butcher: marine: England
LB	979/14	Parke 1950, Plymouth No. 29: type material: marine
LB	979/17	Butcher: marine: England

C. **major** Butcher

LB	979/11	Butcher 1958: type material: marine: England

C. **ovata** Ehrenberg *var.* **palustris** Pringsheim

	979/3	Pringsheim: type material: Czechoslovakia

C. **pseudobaltica** Butcher

LB	979/9	Butcher 1961: type material: marine: Germany

C. **reticulata** Lucas

LB	979/15	Jowett 1965, Plymouth No. 358: marine: England

C. **spp.**

LB	979/18	Plymouth No. 65: marine

The collection has many freshwater strains, none axenic, which were isolated by Pringsheim while in Cambridge. Strain 37 *(C. curvata ?)* is normally supplied for demonstration purposes.

CRYSTALLOLITHUS Gaarder et Markali see *Coccolithus*.

CYANIDIUM Geitler

C. caldarium Geitler *em.* Hirose
1355/1 Allen, No. C14.1.1

CYANOPHORA Korshikov

C. paradoxa Korshikov
LB 981/1 **Pringsheim**

CYATHOMONAS Fromentel

C, truncata Ehrenberg
LB 982/1a **Pringsheim 1942: England**
LB 982/1b **Schuster 1963: California USA**

CYCLIDIOPSIS Korshikov see *Astasia*

CYCLIDIUM Müller

C. glaucoma Müller
LB 1616/1 **George 1962: Russia**

CYCLOTELLA Kützing

C. cryptica Reimann, J. Lewin *et* Guillard
1070/1 **from Göttingen collection: brackish: type material**

CYLINDROCAPSA Reinsch

C. involuta Reinsch
LB 314/1 **George 1950: England. Hoffman and Hofmann (1975) doubt the identify of this strain**

CYLINDROCYSTIS Meneghini

C. brebissonii Meneghini
LB 615/1a **George 1948: England**

CYLINDROSPERMUM Kützing

C. *aff.* **alatosporum**
B 1415/3 **Wilcox 1971: England**

C. maius Kütz.
B 1415/2 **Komarek/Baker 1961: N5: Czechoslovakia**

C. sp.
LB 1415/1 Fogg 1945

CYSTOCOCCUS Naegeli see *Trebouxia*

DACTYLOCOCCUS Naegeli

D. bicaudatus A. Br.

	223/1	Flint 1959: soil: New Zealand

DERBESIA Solier

D. ? tenuissima

LB	706/1	Starr 1959, IU 1260: marine: Mediterranean

DERMOCARPA Crouan

D. violacea Crouan

	1416/1	Lewin 1965: marine: California USA

DIACRONEMA Prauser

D. vlkianum Prauser

LB	914/1	Butcher/Parke 1961, Plymouth No. 244: marine: England
LB	914/2	Hibberd 1975: England

DICHOTOMOSIPHON Ernst

D. tuberosus (A.Br.) Ernst

LB	707/1	Korn 1960: Indiana No. 1036: Indiana USA

DICRATERIA Parke

D. inornata Parke

LB	915/1	Gross 1935/36: Plymouth B: type material: marine: England

DICTYOCHLORIS Vischer *ex* Starr

D. fragrans Vischer *ex* Starr

	220/1	Vischer 1942, No. 334: type material (Starr 1955) from soil: Switzerland
	249/3	Pringsheim 1940: material from John as *Muriella magna*: soil: Engla

DICTYOCOCCUS Gerneck, see also *Bracteacoccus*

D. varians Gerneck *em.* Starr

LB	221/5	Starr 1951: type material: Scotland

DICTYOSPHAERIUM Naegeli

D. ehrenbergianum Naegeli

LB	221/1c	Jaworski 1966, L141: England

D. pulchellum Wood

	222/1a	Pringsheim 1940: England
	222/2a	George 1949: France
	222/2b	Lewin 1952: Nova Scotia, Canada
B	222/2c	Bucka: Poland
LB	222/2d	Jaworski 1972, FBA L246: England

DICTYOSTELIUM Brefeld

D. discoideum Raper
B 1515/2 Ashworth 1969: England

D. mucoroides Brefeld
B 1515/1 Page 1964: Idaho USA

DILABIFILUM Tschermak-Woess

D. arthopyreniae (Vischer *et* Klement) Tsch.-Woess
B 415/2 Tschermak-Woess

D. incrustans (Vischer) Tschermak-Woess
B 415/1 Tschermak-Woess 1970: *ex Verrucaria aquatilis* Mudd: Austria

D. printzii (Vischer) Tschermak-Woess
467/1 Vischer 1926, No. 10: type material: Switzerland

DIMORPHOCOCCUS A. Braun

D. lunatus A. Braun
224/1 George 1949: from Paris No. 90

DIPLOSPHAERA Bialosuknia *em.* Vischer

D. sp..
416/1 Vischer No. 556

DISCOPHRYA Lachmann

D. collini (Root)
LB 1618/2 Paulin and Cook 1967: Georgia USA

DISTIGMA Ehrenberg

D. curvata Pringsheim *var.* **major** Pringsheim
LB 1216/1 Pringsheim 1935: type material: Czechoslovakia

D. gracilis Pringsheim
LB 1216/2 Pringsheim 1936: type material: Austria

D. proteus Ehrenberg
LB 1216/3a Pringsheim: Czechoslovakia
LB 1216/3b Pringsheim: England
LB 1216/3c Pringsheim 1948: England

D. senii Pringsheim
LB 1216/4 Pringsheim 1940: type material: England

DRAPARNALDIA Bory

D. plumosa (Vaucher) Agardh
418/1a Reynolds: Wales
This strain does not satisfactorily show the characterisitic main axis of plants of this genus

DUNALIELLA Teodoresco

D. bioculata Butcher
LB 19/4 Mainx: Prague 281a: type material: Russia
Listed as *D. marina* in the Göttingen and Prague collections.

D. minuta Lerche
LB 19/5 Jowett 1967, Plymouth No. 430: marine: France

D. parva Lerche
LB 19/9 Butcher 1956: marine: England

D. parva Lerche *f.* **eugametos** Lerche
LB 19/1 Pringsheim: material from Lerche

D. percei Nicolai
LB 19/2 Pringsheim 1935: material from Nicolai: reputed type material but this strain does not closely resemble the published description

D. polymorpha Butcher
LB 19/7a Butcher 1960: marine: England
LB 19/7b Butcher 1959: marine: England
LB 19/7c Butcher 1954: marine: England

D. primolecta Butcher
11/34 George 1950, from Gross's Plymouth No. 81, long known as 'the Plymouth Chlamydomonas': type material: marine: England

D. quartiolecta Butcher
LB 19/8 Butcher 1953: marine: England

D. salina (Dunal) Teodoresco
LB 19/3 Mainx, material from Pascher

D. tertiolecta Butcher
19/6a Foyn 1928/Cann 1975, Plymouth No. 83, previously *Chlamydomonas III*: type material: marine: Norway
19/6b Plymouth No. 83: Indiana No. 999: treated with antibiotics by Overnell

DYSMORPHOCOCCUS Takeda

D. globosus Bold *et* Starr
LB 20/1 Bold 1951: type material: Tennessee USA

ECHINAMOEBA Page

E. exundans (Page)
B 1534/4 Page 1965: FCP No. 46: type material: Alabama USA

E. silvestris Page
B 1519/1 Page 1973: FCP No. 154: type material: soil: England

ECHINOSPHAERIUM

E. nucleofilum (Barrett)
LB 1507/1 originally from Carolina Biological Company as *Actinosphaerium*.

ECTOCARPUS Lyngbye

E. siliculosus Harvey

LB	1310/2	Russell 1970

E. ? variabilis Vickers

LB	1310/1	Lewin: marine: California USA

ELAKATOTHRIX Wille

E. viridis (Snow) Printz, = *Fusola viridis* Snow

	227/1	George 1949, from Paris No. 38

ELLIPSOIDION Pascher

E. acuminatum Pascher

LB	822/1	Westlake/Trown 1969: England

EMILIANIA Hay *et* Mohler

E. huxleyi (Lohm.) Hay *et* Mohler

LB	920/1a	Parke 1950, Plymouth No. 92: marine: England
LB	920/1c	Parkins 1971, Plymouth No. 92c: marine: England
		Neither strain produces coccoliths

ENTOPHYSALIS Kützing

E. rivularis (Kütz.) Drouet

LB	1422/1	Norby, Indiana No. 131

ENTOSIPHON Stein

E. sulcatum (Duj.) Stein

LB	1220/1a	George 1965: England
LB	1220/1b	George 1966; England

EREMOSPHAERA De Bary

E. gigas (Archer) Fott *et* Kalina (= *Oocystis eremosphaeria)*

LB	257/4	Staff 1952: Indiana USA

E. viridis De Bary

LB	228/1b	Whybrow: England

EUASTRUM Ralfs

E. verrucosum (Ehrenberg) Ralfs

LB	624/1	Ott 1961: Virginia USA

EUDORINA Ehrenberg

E. charkowiensis (Korshikov) Pascher (= *Pandorina charkowiensis* Korsh.)

	24/2a	Droop 1951: Finland
	24/2b	Peck 1962 from Ott's 068: Virginia USA

E. elegans Ehrenberg

	24/1a	Mainx, Strain 1: female: Czechoslovakia
	24/1b	Rodhe, No. 1637b: Sweden

E. illinoisensis (Kofoid) Pascher (= *Pleodorina illinoisensis* Kof.)

	162/2a	Stein 1956: male: Minnesota USA
	162/2b	Stein 1956: female: Minnesota USA

E. unicocca Smith

	24/1c	Starr, Indiana No. 737: female
	24/1d	Starr, Indiana No. 738: male

EUGLENA Ehrenberg

E. acus *var.* **gracilis** Pringsheim

LB	1224/1d	Pringsheim 1943: type material: England

E. acus *var.* **major** Pringsheim

LB	1224/1b	Pringsheim: type material: England

E. agilis Carter see *E. pisciformis* Klebs

E. anabaena Mainx

	1224/15b	Pringsheim 1940: England
	1224/15c	Pringsheim 1941: England

E. anabaena *var.* **minor** Mainx

	1224/2	Mainx 1924: type material: Czechoslovakia
LB	1224/15d	Pringsheim 1943: England

E. cantabrica Pringsheim

LB	1224/33	Pringsheim 1943: type material: England

E. caudata Hübner

LB	1224/24b	Pringsheim: England

E. clara Skuja

LB	1224/27	Pringsheim 1949: England

E. communis Gojdics

LB	1224/35	Leedale 1957: Scotland

E. cuneata Pringsheim

LB	1224/32a	Pringsheim 1944: type material: England
LB	1224/32b	Pringsheim 1945: type material: England

E. deses Ehrenberg

LB	1224/19a	Pringsheim 1940: England
LB	1224/19b	Pringsheim 1943: England
LB	1224/19c	Pringsheim 1943: England
LB	1224/19d	Pringsheim as 'No. 3, *var intermedia?*'
LB	1224/20	Dusi: type material of *E. mesnili* Deflandre & Dusi
LB	1224/3	Pringsheim 1941

E. deses *var.* **vermiformis** Carter

LB	1224/22	Pringsheim 1951: from brackish water: England

E. ehrenbergii Klebs

LB	1224/36	Pringsheim: Germany

E. geniculata Dujardin

LB	1224/4e	Pringsheim 1942: England
LB	1224/4f	Pringsheim 1951: Austria

E. geniculata *var.* **terricola** Dangeard

	1224/4b	Mainx 1923 (as *E. deses*): Czechoslovakia
	1224/4c	Vischer, No. 400: Switzerland
LB	1224/4d	Pringsheim 1943: England
LB	1224/40a	Pringsheim as strain 2 } These two strains are of **doubtful** identity;
LB	1224/40b	Pringsheim as strain 3 } in some respects they **resemble** ***E. viridis***

E. gracilis Klebs

	1224/5a	Mainx 1927 (1)*: Czechoslovakia
	1224/5b	Elmore-Sauer 1935 (2)
	1224/5c	Vischer 1936, No. 22 (3)
	1224/5d	Pringsheim 1950 (26) material from Sister Monica **Taylor**
	1224/5e	Pringsheim 1940 (5) material from Fogg: England
	1224/5f	Pringsheim 1940 (6) material from Harris: England
	1224/5g	Pringsheim 1939 (4) material from Sister Monica **Taylor: gave** rise to *Astasia longa* 1204/17d **(see Pringsheim 1948)**
	1224/5i	Pringsheim 1948 (19) from Paris No. 101
	1224/5k	Pringsheim 1945 (11) material from Chu
	1224/5l	Pringsheim 1945 (10) material from Mackinnon
	1224/5m	Pringsheim 1944 (8) material from Sister Monica **Taylor**
	1224/5n	Pringsheim 1948 (17): England
	1224/5q	Provasoli 1948, No. 7 (16)
	1224/5s	Pringsheim 1948 (18): England
	1224/5u	Hartshorne 1950 (22): England
	1224/5v	Pringsheim 1948 (20) from Paris No. 115
	1224/5w	Pringsheim 1948 (21) from soil: England
	1224/5x	Pringsheim 1949 (23) France
	1224/5y	Pringsheim 1949 (24): England
	1224/5z	Pringsheim 1950 (25) material from de Saedeleer; **the well-known** **'Z' strain, widely used for the bioassay of vitamin** B_{12}

E. gracilis *var.* **saccharophila**

	1224/5h	Lackey (7)*: gave rise to *Astasia longa* 1204/17e **(see Pringsheim** 1948)
	1224/5r	Pringsheim 1946 (13): England
	1224/5t	Pringsheim 1945 (9) material from Tomlinson: England
	1224/6	Provasoli 1948 (14) *'var. urophora'*
	1224/7a	Cori 1938 (15) *'var. bacillaris'* used for vitamin B_{12} **bioassay**
	1224/7b	a colourless mutant of 1224/7a induced by **streptomycin, designated** SM-L1 by Gross and Jahn

E. granulata (Klebs) Schmitz

	1224/8b	Provasoli: type material of *E. rostrifera* Johnson
LB	1224/8c	Pringsheim: England

E. laciniata Pringsheim

LB	1224/31	Pringsheim 1951: type material: Austria

*These numbers against strains of *E. gracilis* refer to Pringsheim 1952.

E. limnophila Lemmermann
LB 1224/23 Pringsheim: England

E. magnifica Pringsheim
LB 1224/30 Pringsheim 1945: type material: England

E. mesnili Deflandre *et* Dusi see *E. deses* 1224/20

E. mutabilis Schmitz
1224/9a Pringsheim as *E. deses* 8, material from Chu: England
1224/9b Mainx 1924 as *E. klebsii* (Lemm.) Mainx: Czechoslovakia
1224/9c Lewin: Canada
LB 1224/9d Pringsheim 1943: England

E. pisciformis Klebs
LB 1224/18f Pringsheim 1949: England
LB 1224/18g Pringsheim 1948: England
LB 1224/18h Pringsheim 1948: (? = *var. minima* Mainx): England
LB 1224/39 Pringsheim 1947 (of doubtful identity): England

E. pisciformis *var.* **fallax** Pringsheim
LB 1224/18c Pringsheim 1943: England

E. pisciformis *var.* **lata** Pringsheim
LB 1224/18d Pringsheim 1943: England

E. pisciformis *var.* **procera** Pringsheim
LB 1224/18b Pringsheim 1951: France

E. pisciformis *var.* **striata** Pringsheim
LB 1224/18e Pringsheim 1945: England

E. polymorpha Dangeard
LB 1224/26 Pringsheim 1938: Czechoslovakia

E. schmitzii Gojdics = *E. geniculata q.v.*

E. sociabilis Dangeard
LB 1224/12a Pringsheim 1938: Czechoslovakia
LB 1224/12b Pringsheim 1940: England

E. spirogyra Ehrenberg
LB 1224/13a Pringsheim 1938: Czechoslovakia
LB 1224/13b Pringsheim 1940: England
LB 1224/13c Pringsheim 1947: England

E. spirogyra *var.* **fusca** Klebs
LB 1224/37 Leedale: Scotland

E. splendens Dangeard
LB 1224/29a Pringsheim 1944: England

E. stellata Mainx
1224/14 Mainx: type material: Czechoslovakia

E. myxocylindracea Bold *et* MacEntee *var.* **terricola** Bold *et* MacEntee
1224/21 Lewin 1950: Indiana No. 457: type material: Connecticut USA

E. tripteris (Dujardin) Klebs

LB	1224/16a	Pringsheim 1936: Austria
LB	1224/16b	Pringsheim 1943: England

E. tristella Chu

LB	1224/34	Pringsheim

E. velata Klebs

LB	1224/25	Pringsheim 1944: England

E. viridis Ehrenberg

	1224/17a	Pringsheim 1940: a large form: England
	1224/17b	Pringsheim: from Paris No. 108
	1224/17c	Pringsheim 1951: England
LB	1224/17f	Pringsheim 1940: England
LB	1224/17g	Pringsheim 1941: a large form: England
LB	1224/17h	Pringsheim 1943: England
LB	1224/17k	Pringsheim 1944: England
LB	1224/17l	Pringsheim 1949: a small, very metabolic form: England
LB	1224/17m	Pringsheim 1949: England
LB	1224/17n	Pringsheim 1949: England
LB	1224/17p	Pringsheim 1950: England
LB	1224/17q	Pringsheim 1949: England

E. viridis *var.* **halophila** Pringsheim

LB	1224/17e	Pringsheim 1951: type material: brackish: England

E. viridis *var.* **maritima** Pringsheim

LB	1224/17d	Pringsheim 1951: type material: brackish: England

E. sp.

	1224/38	Leedale as '5': a vigorous strain resembling *E. gracilis*

EUGLYPHA Dujardin

E. acanthophora Ehrenberg

LB	1520/3	Hedley 1972: sewage: England

E. rotunda Wailes

LB	1520/1	Hedley 1968: England

E. strigosa (Ehrenberg)

LB	1520/2	Hedley 1970: England

EUPLOTES Ehrenberg

E. parkei Curds

LB	1624/3	Parke 1972: type material: saline: Austria

E. rariseta Curds, West *et* Dorahy

LB	1624/2a	Andrews 1971: marine: England

EUTREPTIA Perty

E. lanowii Steuer
LB 1226/2 Pringsheim 1951: brackish: England

E. pertyi Pringsheim
LB 1226/3 Pringsheim: type material: brackish: England

E. viridis Perty
LB 1226/1a Pringsheim 1947: brackish: England
LB 1226/1c Pringsheim: brackish: England

EUTREPTIELLA Da Cunha

E. spp.
LB 1227/2 Jowett 1965, Plymouth No. 359: marine: England
LB 1227/3 Jowett 1975, Plymouth No. 395: marine: England

E. gymnastica Throndsen
LB 1227/4 Throndsen 1964, Plymouth No. 462: type material: marine: Norway

EXUVIAELLA Cienkowsky see *Prorocentrum*

FIBROCAPSA Toriumi *et* Takano

F. japonica Toriumi *et* Takano
LB 1177/1 Toriumi 1972: type material: marine: Japan

FILAMOEBA Page

F. nolandi Page
B 1526/1 Page 1964: FCP No. 22: type material: Minnesota USA

FISCHERELLA Gomont

F. muscicola (Thuret) Gomont
LB 1427/1 Mitra, No. 6

F. sp.
B 1427/2 Komarek: mangroves: Cuba

FLABELLULA Schaeffer

F. calkinsi (Hogue)
B 1529/1 Page 1969: FCP No. 80: marine: Maine USA

F. citata Schaeffer
B 1529/2 Page 1969: FCP No. 90: marine: Maine USA

FRAGILARIA Lyngbye

F. crotonensis Kitton
LB 1029/1 Jaworski 1974, FBA L237: England

FREMYELLA J. de Toni

F. **diplosiphon** (Bornet *et* Flahault) Drouet
B 1429/1 Strout

FUSOLA Snow

F. **viridis** Snow
227/1 George 1949 from Paris No. 38, as *Elakatothrix viridis*

GEMINELLA Turpin

G. **sp.**
LB 333/1 Lund 1958, FBA L149: England

GLAUCOCYSTIS Itzigsohn

G. **nostochinearum** Itzigsohn
229/1 George/Lewin 1963: England

GLAUCOSPHAERA Korshikov

G. **vacuolata** Korsh.
LB 130/1 Starr, University of Indiana No. 1662

GLENODINIUM Ehrenberg

G. **foliaceum** Stein
LB 1116/3 Ott, Va 43: marine: Virginia USA

G. **sp.**
LB 1116/2 Ott, Va 32: marine: Virginia USA

GLOEOCAPSA Kützing

G. **alpicola** (Lyngbye) Bornet
B 1430/1 Allen, identified by Drouet

G. **spp.**
L 1430/2 M. M. Allen: Stanier 6501: California USA
1430/3 Markle: Stanier 6909: University of Indiana No. 795

GLOEOCOCCUS A. Braun

G. **maximus** (Mainx) Fott *et* Novakova
31/1 Mainx 1925: type material: Czechoslovakia

GLOEOCYSTIS Naegeli

G. **maxima** Mainx = *Gloeococcus maximus* (Mainx) Fott *et* Novakova

G. **vesiculosa** Naegeli
31/3 Lewin 1950 (not the same as Göttingen 31/3): Connecticut USA

GLOEODINIUM Klebs

G. montanum Klebs
B 1120/1 von Stosch: terrestrial: Germany

GLOEOMONAS Klebs

G. kupfferi (Skuja) Gerloff
LB 33/1 Bourrelly, Paris No. 36

GLOEOTRICHIA Agardh

G. echinulata (Smith) Richter
LB 1432/1 George 1950: Sweden

GOLENKINIOPSIS Korshikov

G. parvula (Woronich) Korshikov
231/1 George 1949: England

GOMONTIA Bornet *et* Flahault

? G. sp.
LB 432/1 Lewin Y1/12: does not penetrate shell: marine: Canada

GOMPHONEMA Agardh

G. parvulum Kützing
B 1032/1 J. Lewin 1949: Connecticut USA

GONATOZYGON de Bary

G. monotaenium de Bary
LB 632/1 King 1963: Wales

GONGROSIRA Kützing

G. terricola Bristol
434/1 Flint/King: soil: New Zealand

GONIUM Müller

G. pectorale Müller
32/1a Pringsheim 1940: England
32/1b Droop 1961: Finland (not the same as Göttingen 32/1b)
32/1c Hibberd 1972: England

G. sociale (Dujardin) Warming

	32/2a	Pringsheim 1941: England
	32/2b	Starr 1951: homothallic: England
	32/3	Meyer: formerly listed as *G. tetras*: Germany

GONYAULAX Diesing *em.* Kofoid

G. tamarensis Lebour

LB	1119/1	Adams 1957/71, Plymouth No. 173: marine: England

G. diegensis Kofoid

LB	1119/2	Ott

GYMNODINIUM Stein

G. pigmentosum (Dodge) Loeblich *(Aureodinium)*

LB	1103/1	Parke 1958, Plymouth No. 208: type material: marine: England

G. *aff.* **simplex** (Lohm.) Kofoid *et* Swezy

LB	1117/3	Jowett 1965, Plymouth No. 368: marine: England

GYROPAIGNE Skuja

G. lefevrei Bourrelly *et* Georges

LB	1233/1	Christen

HAEMATOCOCCUS Flotow

H. capensis Pocock

LB	34/4b	George/Pringsheim: material from Pocock: South Africa

H. droebakensis Wollenweber

	34/2a	Pringsheim 1950: Sweden
	34/2c	Pringsheim 1950: Sweden
	34/2g	George 1950: material from Pocock: England

H. droebakensis *var.* **fastigata** Wollenweber

LB	34/3	Droop 1951: Finland

H. lacustris (Girod-Chantrans) Rostafinski

	34/1a	Pringsheim: material from Naumann
	34/1b	Pringsheim: Czechoslovakia
	34/1c	Mainx: Czechoslovakia
	34/1d	Vischer 1923, No. 2: Switzerland
	34/1e	George 1950: England
	34/1f	Pringsheim 1951: Spitzbergen
	34/1h	Droop 1951: Finland
	34/1j	Lewin 1951: England

H. pluvialis Flotow *em.* Wille = *H. lacustris q.v.*

H. zimbabwiensis Pocock

L	34/5	George 1964: material from Pocock: South Africa

HALOCHLORELLA Dangeard

H. rubescens Dangeard = *Chlorella fusca var. rubescens* Kessler *et al.*
232/1 Izard 1966: type material: France

HALOCHLOROCOCCUM Dangeard

H. marinum Dangeard
B 233/1 Izard 1965: type material: marine: France

HALOSPHAERA Schmitz

H. minor Ostenfeld *in* Knudsen *et* Ostenf.
LB 135/3 Plymouth No. 205c: marine

H. russellii Parke
LB 135/1 Adams 1961, Plymouth No. 247: type material: marine: England
Both strains are in flagellate stage only

HARTMANNELLA Alexeieff (see *Acanthamoeba*)

H. cantabrigiensis Page
B 1534/8 Page 1972, No. 123: type material: England

H. castellanii Douglas see *Acanthamoeba*

H. limacoides see *Cashia*

H. limax see *Saccamoeba*

H. rhysodes Singh see *Acanthamoeba*

H. vermiformis Page
B 1534/7 Page, No. 25: type material: Wisconsin USA

HEMISELMIS Parke

H. brunnescens Butcher
LB 984/2 Parke 1949, Plymouth No. 14: type material: marine
LB 984/6 ? Butcher: marine

H. rotunda Butcher
LB 984/4 Butcher: marine

H. rufescens Parke
LB 984/1 Parke 1937, Plymouth 'D': type material: marine: Isle of Man

H. virescens Droop
LB 984/5 Adams 1956, Plymouth No. 157: marine

HETERAMOEBA Droop

H. clara Droop
B 1536/1 Droop 1960: type material: marine: Scotland

HETEROCAPSA Stein

H. triquetra (Ehr.) Stein

LB	1125/1	Adams 1957, Plymouth 169: marine: England
LB	1125/2	Ott 1975, Va 14: marine: Virginia USA

HETEROCOCCUS Chodat

H. brevicellularis Vischer

	835/1	Vischer 1943, No. 351: from soil: Switzerland

H. caespitosus Vischer

	835/2a	Vischer 1934, No. 116: from soil: Germany
B	835/2b	Vischer 1934, No. 131: Switzerland

H. chodatii Vischer (*H. viridis* Chodat)

	835/3	Chodat: Vischer No. 161

H. crassulus Vischer

	835/4	Vischer 1943, No. 367: Switzerland

H. fuornensis Vischer

	835/5	Vischer No. 279: Switzerland

H. mainxii Vischer

	835/6	Mainx 1926: Vischer No. 160: from soil: Czechoslovakia

H. marietanii Vischer

	835/7	Vischer 1936, No. 167: Switzerland

H. protonematoides Vischer

	835/9	Vischer 1945, No. 369: Switzerland

HETEROGONIUM Dangeard

H. salinum Dangeard

	234/1	Lepaileur: marine: France

HETEROMASTIX Korshikov

H. ? angulata

LB	1960/4	Belcher and Swale 1975: England

H. longifilis (Butcher) Rayns

LB	1960/3	Parke 1950, Plymouth No. 58: type material: marine: England

H. rotunda (Carter) Manton

LB	1960/1	Butcher: marine

H. sp.

	1960/2	marine: origin unknown: rendered axenic by Pennick 1965

HETEROTETRACYSTIS Cox *et* Deason

H. macrogranulosa Cox *et* Deason

	136/1	Cox 1964: soil: Tennessee USA

HETEROTHRIX Pascher

H. debilis Vischer
836/1 Vischer 1929, No. 50: Switzerland

H. hormidioides Vischer
836/2 Vischer 1943, No. 358: from soil: Switzerland

H. montana Vischer
836/3 Vischer 1945, No. 288: from soil: Switzerland

H. solida Vischer
836/4 Vischer 1940, No. 214: Switzerland

H. sp.
836/5 Vischer, No. 391

HILDENBRANDIA Nardo

H. rivularis (Lieben) J. Agardh
LB 1368/1 Ott, No. O 341

HORMIDIUM Klebs see *Klebsormidium*

HYALODISCUS Hertwig *et* Lesser see *Cochliopodium*

HYALOPHACUS Pringsheim

H. ocellatus Pringsheim
LB 1237/1a Pringsheim 1947: England
LB 1237/1b Christen: Switzerland

HYALORAPHIDIUM Pascher *et* Korshikov

H. curvatum Korshikov
235/1 Lewin 1949: Connecticut USA

HYALOTHECA Ehrenberg

H. dissiliens (Smith) Brébisson
LB 637/1 Pringsheim 1947: England

HYDRODICTYON Roth

H. africanum Yamanouchi
LB 236/2 George 1948: South Africa

H. reticulatum (L.) Lagerheim
LB 236/1a George 1947: England
LB 236/1b Pirson: Germany
LB 236/1c Marchant 1969: Australia

HYMENOMONAS Stein

H. carterae (Braarud *et* Fag.) = *Cricosphaera carterae q.v.*

H. roseola Stein

LB	925/1	Pringsheim: England

INTERFILUM Chodat

I. paradoxum Chodat

	338/1	Pringsheim: material from Miss John: from soil: England

ISOCHRYSIS Parke

I. galbana Parke

LB	927/1	Parke 1938, Plymouth I: type material: marine: Isle of Man

I. spp.

LB	927/2	Butcher: grows well: marine: England
LB	927/3	Butcher: grows well: marine: England
LB	927/4	Butcher: marine: England
LB	927/5	Butcher 1960: marine: England
LB	927/6	Butcher 1959: marine: England
LB	927/7	Butcher 1956: marine: England
LB	927/8	Butcher 1960: marine: England

I. spp. ?

LB	927/9	Butcher 1959: marine: Wales
LB	927/10	Butcher 1960: marine: England

KENTROSPHAERA Borzi

K. sp..

LB	241/1	Starr 1951: Scotland

KHAWKINEA Jahn *et* McKibben see under *Astasia*

KIRCHNERIELLA Schmidle

K. contorta (Schmidle) Bohlin

	243/3	Wurtz: France

K. lunaris (Kirchner) Moebius

	243/1	Pringsheim 1939: England

K. lunaris *var.* **dianae** Bohlin

	243/4	Wurtz: France

KLEBSORMIDIUM Silva, Mattox *et* Blackwell (= *Hormidium* Klebs)

K. flaccidum (Kütz.) S.M. *et* B.

	335/1a	Pringsheim: material from Barlow: (Wisconsin USA?)
	335/1b	Pringsheim: material from Barlow: (Wisconsin USA?)
	335/2a	Pringsheim: material from Barlow: (Wisconsin USA?)
	335/2b	Pringsheim: material from Barlow: (Wisconsin USA?)
	335/4	Pringsheim: as ? *H. stoechidum*
	335/9	Lewin 1952: Nova Scotia Canada

K. subtilissimum (Rabenhorst) S.M. *et* B.

384/1 Lewin 1952: from snow: Alaska

The following strains, formerly identified as *Hormidium* probably belong here.

LB	335/6	Pringsheim as *Hormidium crenulatum* Kützing: material from John: England
	335/3	Pringsheim as *Hormidium nitens*
	335/5	Strout 1951: Connecticut USA
	335/7	Algéus, No. 1: Sweden
B	335/8	Czurda: Czechoslovakia
	335/9	Lewin 1952: Nova Scotia Canada

LAGERHEIMIA Chodat

L. genevensis Chodat

LB 246/1 Belcher and Swale 1975: England

LAMBORNELLA Keilin = *Tetrahymena q.v.*

LAMPROPEDIA Schroeter

L. hyalina (Ehrenberg) Schroeter

1444/1 Pringsheim 1951 (Göttingen 1740/1): England

LAUTERBORNIA Pringsheim see under *Synechococcus leopoliensis*

LEPTOMYXA Goodey

L. reticulata Goodey

B 1546/1 Page 1971: England

LEPTOSIRA Borzi

L. obovata Vischer = *Pleurastrum obovata* (Vischer) Tupa

445/1 Vischer 1928, No. 46: type material: Switzerland

LOBOMONAS Dangeard

L. piriformis Pringsheim

45/1 Pringsheim 1930: type material: Czechoslovakia

LYNGBYA Agardh *ex* Gomont

L. majuscula Harvey

LB 1446/4 George 1953: brackish: England

L. spp.

	1459/2	Manten: Utrecht P. 35, formerly listed as *Oscillatoria*
B	1446/5	Butcher: marine

MALLOMONAS Perty

M. cratis Harris *et* Bradley
LB 929/1 Belcher 1962: England

M. papillosa Harris *et* Bradley
LB 929/2 Belcher 1965: England

MANTONIELLA Desikachary

M. squamata (Manton *et* Parke) Desikachary
LB 1965/1 George 1951: type material: brackish: England
LB 1965/5 Butcher, as *Thalassomonas caeca*: marine

MASTIGOCLADUS Kirchner

M. laminosus Cohn
1447/1 Fogg: from hot springs: New Zealand

MAYORELLA Schaeffer

M. palestinensis Reich see *Acanthamoeba*

M. riparia Page
B 1547/3 Page 1971: type material: England

M. viridis Leidy
LB 1547/4 George 1966: England

MELOSIRA Agardh

M. italica (Ehr.) Kützing
LB 1048/3 Jaworski 1965, L154: England

M. varians Ag.
LB 1048/1 Belcher 1974: England
LB 1048/2 Lund 1970, FBA L222: England

MENOIDIUM Perty

M. bibacillatum Pringsheim
LB 1247/1 Pringsheim 1940: type material: England

M. cultellus Pringsheim
LB 1247/2 Pringsheim: type material: Czechoslovakia

M. intermedium Pringsheim
LB 1247/3 Pringsheim 1940: type material: England

M. obtusum Pringsheim
LB 1247/4 Pringsheim: Germany

M. sp.
LB 1247/6 Pringsheim as 'X'

MERISMOPEDIA Meyen

M. convoluta Brébisson
B 1448/3 Pringsheim 1967: Germany

M. glauca (Ehr.) Naegeli *f.* **insignis** (Schkorb.) Geitler
LB 1448/1 Pringsheim 1947: England

M. punctata Meyen
LB 1448/2 Komarek

MESOSTIGMA Lauterborn

M. viride Lauterborn
LB 50/1 Pringsheim 1943: England

MESOTAENIUM Naegeli

M. caldariorum (Lagerheim) Hansgirg
230/1 van Overbeek: HMS 2.5.1.1
648/1 Czurda 1924: Czechoslovakia

MICRACTINIUM Fresenius

M. pusillum Fresenius
248/1 George 1954: England

MICRASTERIAS Agardh

M. papillifera Brébisson
LB 649/7 Lefevre 1936

M. rotata (Greville) Ralfs
LB 649/4b Kallio: diploid strain
LB 649/9 Hibberd 1970

M. thomasiana *var.* **notata** Gronblat
LB 649/5 Kallio: haploid and diploid

M. truncata (Corda) Brébisson
LB 649/6 King 1952: England

MICROCHAETE Bornet *et* Flahault

M. grisea Thuret *ex* Bornet *et* Flahault
B 1445/1 Butcher: marine: England

MICROCOLEUS Gomont

M. paludosus (Kützing) Gomont
B 1449/1 George 1950: England

M. vaginatus (Vaucher) Gomont *var.* **radiatus** Baker *et* Bold
1459/4 Manten Utrecht P. 27: soil: Holland. Identified by Baker and Bold

MICROCYSTIS Lemmermann

M. aeruginosa Kützing (= *Diplocystis aeruginosa* (Kütz.) Trev.)
LB 1450/1 Gerloff 1948, No. 1036: Wisconsin USA

M. incerta Lemmermann
B 1450/2 from Griefswald via Trebon

MICROMONAS Manton *et* Parke

M. pusilla (Butcher) Manton *et* Parke
LB 1965/4 Parke 1950, Plymouth No. 27: marine

M. squamata Manton *et* Parke see *Mantoniella*

MICROSPORA Thuret

M. amoena (Kütz.) Rabenhorst
LB 348/1 George 1948: England

MICROTHAMNION Naegeli

M. kutzingianum Naegeli
B 450/1a Christensen 1949: England
450/1b Pringsheim 1948: England

MISCHOCOCCUS Naegeli

M. sphaerocephalus Vischer
847/1 Vischer 1929, No. 61: type material: Switzerland

MONAS O.F. Müller

M. pudica Pringsheim
LB 930/1 Pringsheim 1944: England

M. sociabilis Meyer = *Anthophysa vegetans q.v.*

MONOCHRYSIS Skuja

M. lutheri Droop = *Pavlova lutheri q.v.*

MONODUS Chodat

M. subterraneus Petersen
848/1 Lewin: USA

MONOMASTIX Scherffel

M. minuta Skuja
LB 1963/2 Belcher 1967: England

MOUGEOTIA Agardh

M. sp.

LB	650/1	George 1949: cells 30-37μ br: England

MURIELLA Petersen

M. aurantiaca Vischer

	249/1	Vischer 1933, No. 108: terrestrial: Switzerland

M. decolor Vischer

	249/2	Vischer 1926, No. 29: Switzerland

M. magna Fritsch *et* John = *Dictyochloris fragrans q.v.*

MYRMECIA Printz

M. pyriformis Tschermak-Woess *et* Plessl

LB	250/1	Tschermak-Woess: type material: from gonidium of *Biatorella simplex*: Austria

M. reticulata Tschermak-Woess

	250/3	Tschermak-Woess *ex Phlyctis argena*

M. sp.

	250/2	Ahmadjian 1965 *ex Dermatocarpon tuckmanii*

MYXOSARCINA Printz

M. chroococcoides Geitler

	1451/1	Greifswald No. A149

NAEGLERIA Alexeieff

N. fowleri Carter

B	1518/3	Jamieson 1972 'Morgan': Australia
B	1518/4	Jamieson 1972 'PA90': Australia
B	1518/5	Butt ca. 1967 'HB1': type of *N. aerobia* Singh *et* Das: USA

These strains are dangerous pathogens and are only issued to recognised laboratories capable of handling them safely.

N. gruberi (Schardinger) Alexeieff

B	1518/1a	Pringsheim, used by Willmer: cysts not typical of species
B	1518/1b	Pringsheim
B	1518/1c	Pringsheim
B	1518/1d	Pringsheim/Fulton
B	1518/1e	Page: Wisconsin USA
B	1518/1f	Page: Indiana USA
B	1518/1g	Page: Alabama USA
B	1518/1s	Pringsheim, used by Singh, probably is one of the Pringsheim strains above

N. jadini Willaert *et* Le Ray

B	1518/2	Jadin: type material: Belgium

NANNOCHLORIS Naumann

N. atomus Butcher

	251/4a	Knight-Jones 1948/Cann 1975: marine: Plymouth 84 as *Ulvella lens:* England
LB	251/4b	Butcher 1960: marine: England

N. coccoides Naumann

	251/1a	Lewin 1949: Connecticut USA
	251/1b	George 1951: England

N. maculata

LB	251/3	Butcher: marine

N. oculata Droop

B	251/6	Ott; Va 19: marine: Virginia USA

N. sarniensis

LB	251/2	Butcher

N. sp.

LB	251/5	Butcher 1959: marine: Wales

NAUTOCOCCUS Korshikov

N. emersus Geitler

	53/4	Bold 1974: Texas, USA

N. piriformis Korshikov

	53/1	Starr: University of Indiana 125: from soil: Massachusetts USA

N. soluta Archibald

B	53/2	Bold: type material: soil: Texas USA

N. terrestris Archibald

B	53/3	Bold: type material: soil: Texas USA

NAVICULA Bory

N. complanatula Hustedt

LB	1050/5*	Belcher 1975: marine: England

N. pelliculosa Brébisson

	1050/3a	Lewin: Connecticut USA

N. sp.

	1050/1	George 1948: from soil: South Africa

NEOBURSARIDIUM Balech

N. gigas Balech

LB	1652/1	Nilsson 1956: Uganda

NEOCHLORIS Starr

N. conjuncta Archibald
254/1 Bold 1968: type material: Texas USA

N. texensis Archibald
B 254/2 Bold 1968: type material: soil: Texas USA

N. vigensis Archibald
254/3 Bold 1968: type material: Texas USA

N. wimmeri (Rabenhorst) Archibald *et* Bold
213/4 Mainx: Meyer Chlor. 7.

NEOSPONGIOCOCCUM Deason

N. **granatum** (Starr) Deason
213/1a Pringsheim: type material: soil

NEPHROCHLAMYS Korshikov

N. subsolitaria (West) Korshikov (*Kirchneriella subsolitaria* West)
B 252/1 Belcher, No. 15
243/2a George 1947: England
243/2b George 1949: Switzerland

NEPHRODIELLA Pascher

N. brevis Vischer
850/1 Vischer 1941, No. 267: from soil: Switzerland

NEPHROSELMIS Stein

N. gilva Parke *et* Raynes
LB 151/1 den Hartog-Adams 1958: Plymouth No. 197: type material: marine: England

N. sp.
LB 151/2 Plymouth No. 204: marine

NETRIUM Itzigsohn *et* Rothe, in Rabenhorst

N. digitus (Ehrenberg) Itzigsohn *et* Rothe
LB 652/1 George 1951: Wales

NITZSCHIA Hassall

N. closterium (Ehrenberg) W. Smith
LB 1052/8 Belcher 1974: marine: England

N. closterium (Ehr.) W. Smith *f.* **minutissima** Allen *et* Nelson
see *Phaeodactylum tricornutum*

N. frustulum Kützing
1052/2 Pringsheim 1948: material from Rice: brackish: Massachusetts USA

N. gotlandica Euler
LB 1052/7 Parke: Plymouth No 121: marine: England

N. kutzingiana Hilse
B 1052/4 Lewin: Massachusetts USA

N. palea (Kützing) W. Smith
1052/3 Harder

N. sigma (Kütz.) W. Smith
LB 1052/10* Belcher 1974: marine: England

N. tryblionella Hantz. *var.* **victoriae** Grun.
LB 1052/9* Belcher 1974: marine: England

NODULARIA Mertens

N. harveyana Thuret
B 1452/1 Butcher: marine

N. spumigena Bornet *et* Flahout
B 1452/4 Nordin c. 1972, No. 1: soil: British Columbia Canada

N. spp.
B 1452/2 Smith and Wilcox 1972 as N1: England
B 1452/3 Smith and Wilcox 1972 as N2: ? *Sphaerocarpa*: England

NOSTOC Vaucher *ex* Bornet *et* Flahault

N. calcicola Brébisson
1453/1 Manten, Utrecht No. P.33 (= *N. ellipsosporum* according to Forest): from soil: Holland

N. commune Vaucher
B 1453/24 du Preez 1971 Mes 1: from *Encephalartos horridus*: South Africa
B 1453/3 Wassink, Utrecht No. P.1: from soil: as *N. punctiforme*
B 1453/29 Donaldson and Whitton, Durham 201: mud: Aldabra Atoll
used for demonstration

N. ellipsosporum (Desmaz.) Bornet *et* Flahault
1453/1 as *N. calcicola* q.v.
1453/2 Manten, Utrecht No. P.42: Holland
1453/11 George 1950 as *N. pruniforme*: Sweden
1453/13 Mitra, No. 3 as *N. paludosum* Kütz.
1453/15 Forest from *Anabaena variabilis* in Drouet's herbarium: Ohio USA
1453/16 Forest from *Anabaena variabilis* in Drouet's herbarium
1453/17 Forest, FOG 470: from Fogg's culture 47b: USA
1453/18 Gerloff, No. 1013j
1453/19 'W' strain of Lazaroff and Vishniac: obtained from Forest as WIS 038

N. entophytum Bornet *et* Flahault

	1453/14	Stewart: marine (= *N. muscorum* according to **Forest**): nitrogen-fixer: Scotland

N. muscorum Agardh

	1453/8	Gibson as *Anabaena* No. 30: Scotland
	1453/9	Gibson as *Anabaena* No. 2: Scotland
	1453/12	Allison, H.M.S. No. Z.56.1.1: nitrogen fixer
	1453/14	Stewart as *N. entophytum* q.v.
	1453/20	Allen No. M.12.4.3: obtained from Forest as **ALN 242**
B	1453/21	Forest, from specimen in Drouet's herbarium
	1453/22	Watanabe 1951: U.L. culture No. 51
	1453/23	du Preez 1971 Mes 7: from *Encephalartos umbeluziensis*: South Africa

N. paludosum Kützing

	1453/13	Mitra, as No. 3 (= *N. ellipsosporum* according to **Forest**)

N. pruniforme Agardh

	1453/11	George 1950 (= *N. ellipsosporum* according to **Forest**)

N. punctiforme (Kützing) Hariot

	1453/3	Wassink, Utrecht P.1: from soil: Holland (= ***N. muscorum*** according to Forest)

N. spp.

	1453/4	Wassink, Utrecht No. P.12: from soil: Holland
	1403/5	Wassink, Utrecht No. P.15: from soil: Holland
	1403/6	Wassink, Utrecht No. P.21: from soil: Holland
	1403/4a	Hecker 1950
B	1453/25	Wilcox and Smith 1971, as Nostoc I: England
B	1453/26	Wilcox and Smith 1971, as Nostoc II; England
B	1453/27	from Rothamsted via Wilcox as '*Nostoc col.*'
B	1453/28	Rodgers 1972: *ex.Anthoceros*: Scotland

OCHROMONAS Wyssotzki (see also *Poterioochromonas*)

O. danica Pringsheim

L	933/2	Pringsheim 1954: type material: Denmark

O. minuta *nom. prov.* Pringsheim

L	933/10	Pringsheim 1951: Germany

O. spp.

L	933/4	Atkinson 1965/Trown 1968: Malaya
LB	933/5	Jowett 1966, Plymouth No. 420: marine: Wales
LB	933/6	Jowett 1963; Plymouth No. 338: marine: England
LB	933/7	Jowett 1963; Plymouth No. 346: marine: England
LB	933/8a	Butcher 1956 as "*adiposus*": marine: England
LB	933/8b	Butcher as "*adiposus*": marine: Wales
LB	933/8c	Butcher as "*adiposus*": marine: England: grows very well
LB	933/9	Butcher 1959 as "*magnoliae*": marine: England
LB	933/17	Butcher 1955 as "*magnoliae*": brackish: England
LB	933/18	Butcher as "*magnoliae*": brackish: England
LB	933/11	Butcher as "*maxima*": marine: England
LB	933/12	Butcher 1954 as "*astigma*": brackish: Wales
LB	933/15	Butcher as "*bicolor*": marine: England
LB	933/16	Butcher as "*bicolor*": marine: England

LB	933/13	Butcher as "*globosa*": brackish: England
LB	933/14	Butcher as "*globosa*": marine: England
LB	933/19	Butcher 1960 as "*phaeocystoides*": brackish: England
LB	933/20	Butcher 1960 as "*phaeocystoides*": brackish: England
LB	933/21	Butcher 1956 as "*stigmatica*": marine: England
LB	933/22	Butcher 1960 as "*stigmatica*": marine: England
LB	933/23	Butcher as "*subglobosa*": marine: England

OEDOCLADIUM Stahl

O. cirratum Beaney *et* Hoffman

LB	574/1	Milliger 1964: Indiana 1532: type material : **Texas USA**

OEDOGONIUM Wittrock

O. cardiacum Wittrock

LB	575/1a	Christensen 1949: male: England
LB	575/1b	Christensen 1949: female: England

O. foveolatum Wittrock

LB	575/2	Bold 1958: homothallic: Indiana No. 933: previously listed as *O. rupestre*

OLISTHODISCUS Carter

O. luteus Carter

LB	934/1	Throndsen 1964, Plymouth No. 461: marine: Norway
LB	934/2	Adams 1959, Plymouth No. 12A: marine: England
LB	934/3	Lackey, Plymouth No. 239: Marine: Florida USA

OOCYSTIS Naegeli

O. apiculata W. West

257/3 Wurtz, No. 24 (= *O. solitaria* Wittrock *f. major* Wille)

O. eremosphaeria G. M. Smith

see *Eremosphaera gigas*

O. marssonii Lemmermann

257/1 Beijerinck: perhaps not *Oocystis sp.*

O. solitaria Wittrock *f.* **major** Wille

257/3 Wurtz No. 24

O. sp.

257/2 Vischer

OPHIOCYTIUM Naegeli

O. majus Naegeli

855/1 Pringsheim: England

OSCILLATORIA Vaucher *ex* Gomont

O. animalis Agardh

LB	1459/6	from University College London

O. lutea Agardh *var.* **contorta** Baker *et* Bold

LB	1459/3	Pringsheim 1941: identified by Baker and Bold

O. princeps Vaucher

LB	1459/7	Aschner 1964: Israel

O. tenuis Agardh

	1459/4	Manten, Utrecht No. P.27: from soil: Holland

O. spp.

B	1459/8	Butcher: brackish
B	1459/9	Butcher: as "*nigro-viridis*": marine
B	1459/13	Butcher: marine: England

OXYRRHIS Dujardin

O. marina Dujardin (with diatom as food)

LB	1133/2	Pennick 1972: marine: England
LB	1133/3	Jowett 1967, Plymouth No. 209b: marine: England
LB	1133/4	Pennick 1974: marine: Bahrain

PALMODICTYON Kützing

P. varium (Naegeli) Lemmermann

LB	59/1	George 1949: seldom achieves typical form in culture: England

PANDORINA Bory

P. charkowiensis Korshikov (= *Eudorina charkowiensis* (Korsh.) Pascher)

24/2a	Droop 1951: Finland
24/2b	Peck 1962: from Ott No. 068: Virginia USA

P. morum Bory

60/1a	Czurda: Czechoslovakia
60/1b	Chu
60/1c	George 1950: England
60/1d	Wilbois, No. In-50-3, + strain: Indiana No. 788
60/1e	Wilbois, No. In-50-11, - strain: Indiana No. 789

PARAMECIUM Hill

P. aurelia Ehrenberg

LB	1660/2a	Pringsheim
LB	1660/2e	George 1965: USA
LB	1660/3	strain GML
LB	1660/3	strain 50
LB	1660/3	strain 51 sensitive
LB	1660/3	strain 51 killer
LB	1660/3	strain J7, Jankowsky 1959: Russia
LB	1660/3	strain J19, Jankowsky 1959: Russia
LB	1660/3	strain J20, Jankowsky 1959, Russia
		(these last two strains have abnormal morphology)

P. bursaria Focke

LB	1660/1a	Pringsheim: Czechoslovakia
LB	1660/1b	George 1949: England
LB	1660/1c	George 1949: England
LB	1660/1d	George 1949: England
LB	1660/1e	Pringsheim: England
LB	1660/1f	Pringsheim: mates with 1660/1g: Scotland
LB	1660/1g	George 1950: mates with 1660/1f: England
LB	1660/1j	Page 1976: mates with 1660/1k: England
LB	1660/1k	Page 1976: mates with 1660/1j: England

P. caudatum Ehrenberg

LB	1660/2c	George 1954: material from H. Gilbert Carter: England
LB	1660/2f	Page 1973: England

P. putrinum Clap. *et* Lach.

LB	1660/7	Patterson 1975: England

P. woodruffi Wenrich

LB	1660/4a	Jankowsky 1959, No. 20: mates with 1660/4b: Russia
LB	1660/4b	Jankowsky 1959, No. 28: mates with 1660/4a: Russia

PARAMOEBA Schaudinn

P. eilhardi Schaudinn

LB	1560/2	Grell 1960: marine: France

P. pemaquidensis Page

B	1560/3	Page 1969: type material: marine: Maine USA

PARAPHYSOMONAS de Saedeleer

P. butcheri Pennick *et* Clarke

LB	935/1	Butcher 1956: type material: cultured with a coccolithophorid: brackish: England

P. vestita (Stokes) de Saedeleer

LB	935/3	Butcher 1960: marine: England
LB	935/4	Butcher: marine: England

P. corbidifera Pennick *et* Clarke

LB	935/2	(in 4/1 *Pyramimonas aff. orientalis*) type material: marine: England

PARMIDIUM Christen

P. scutulum (Skuja) Christen

LB	1258/1	Christen: type material: Switzerland

PASCHERIELLA Korshikov = *Pascherina q.v.*

PASCHERINA Silva

P. tetras (Korshikov) Silva
    LB 159/2 Belcher and Swale 1975: England

PAULSCHULZIA Skuja

P. pseudovolvox (Schulz) Skuja
    58/1 Droop 1951: Finland

P. tenera (Korshikov) Lund (= *P. elegans* Lund)
    LB 58/2 Lund: type material: England

PAVLOVA Butcher

P. gyrans Butcher
    LB 940/1a Butcher 1947: type material: marine: England
    LB 940/1b Knight-Jones 1946: Plymouth No. 93: marine: England
    LB 940/1c Butcher 1957: marine: Wales

P. lutheri (Droop) Green
    LB 931/1 Droop, Millport No. 60: Plymouth No. 75: marine

P. mesolychnon Van der Veer
    LB 940/3 Van der Veer 1967: type material: marine: England

P. pinguis Green
    LB 940/2 Green 1970: type material: marine: Madeira

P. spp.
    LB 931/2 Butcher (as *Monochrysis sp.*): grows well: marine: England
    LB 931/3 Butcher (as *Monochrysis sp.*): marine: England
    LB 931/4 Butcher 1960 (as *Monochrysis sp.*): marine: England

PEDIASTRUM Meyen

P. biradiatum Meyen
    261/1 Rodhe No. 1611a: Sweden

P. boryanum (Turpin) Meneghini *var.* **boryanum**
    LB 261/2 Rodhe

P. boryanum *var.* **cornutum** (Racib.) Salek
    LB 261/3a Pringsheim 1948: England

P. duplex *var.* **reticulatum** Lagerh.
    LB 261/7 Ott No. 201

P. tetras (Ehr.) Ralfs
    261/5 Pringsheim: Czechoslovakia
    261/6 Czurda 1932: Prague No. 216: Austria

PEDINELLA Wysotzki

P. spp.

LB	941/1a	Butcher: as "*marina*": marine: England
LB	941/1b	Butcher: as "*marina*": marine: England
LB	941/2	Butcher: marine: England
LB	941/3	Butcher 1959: marine: England

PEDINOMONAS Korshikov

P. minor Korshikov

B	1965/3b	Hindak 1959: from Ettl as 117b: Czechoslovakia

P. tuberculata (Vischer) Gams

LB	1965/2	Vischer, No. 274: type strain, as *Chlorochytridion tubercuatum:* Switzerland

PELOGLOEA Lauterborn

P. sp.

LB	1461/1	Pringsheim 1948: ? = *Coccochloris peniocystis*: England

PELOMYXA Greeff see *Chaos*

PERANEMA Dujardin

P. trichophorum Stein

	1260/1	Chen 1950, material from Pringsheim

PERIDINIUM Ehrenberg *em.* Stein

P. cinctum (Müller) Ehrenberg

LB	1134/2	Pringsheim, strain 2
LB	1134/3	Pringsheim, strain 1: grows poorly

P. faeroense Paulsen

LB	1134/4	Parke 1957, Plymouth No. 180: marine: England

P. trochoideum (Stein) Lemmermann

LB	1134/1	Parke 1949, Plymouth No. 104: marine: England

PHACOTUS Perty

P. lenticularis Ehrenberg

LB	61/1	Pringsheim: England

PHACUS Dujardin

P. acuminata Klebs

LB	1261/1	Pringsheim 1940: England

P. alata Klebs

LB	1261/2b	Pringsheim 1940: England
LB	1261/2c	Pringsheim 1940: England

P. caudata Hübner

LB	1261/5	Pringsheim 1943: England

P. megalopsis Pochmann

LB	1261/9	Droop 1951: identified by Pochmann: England

P. oscillans Klebs

LB	1261/10	Pringsheim

P. pleuronectes (O.F.M.) Dujardin

LB	1261/3a	Pringsheim 1937: Czechoslovakia
LB	1261/3b	Pringsheim: England

P. pusilla Lemmermann

LB	1261/6	Pringsheim 1945: England

P. pyrum (Ehrenberg) Stein

LB	1261/4a	Pringsheim 1940: England
LB	1261/4b	Pringsheim: England

P. triqueter (Ehrenberg) Dujardin

LB	1261/8	De Bussy 1948: England

PHAEASTER Scherffel

P. pascheri Scherffel

LB	942/1	Belcher 1968: England

PHAEOCYSTIS Lagerheim

P. pouchettii (Hariot) Lagerheim

LB	943/1	Adams 1955, Plymouth No. 147: marine: England
LB	943/2	? 1952, Plymouth No. 64: marine: England

P. spp.

LB	943/3	Butcher 1960: marine: England
LB	943/4	Butcher : marine: Adriatic

PHAEODACTYLUM Bohlin

P. tricornutum Bohlin

	1052/1a	Pringsheim, from the Plymouth strain No. 100, Allen 1910, long known as *Nitzschia closterium f. minutissima*: marine: England
	1052/1b	Barker: H.M.S. Z.9.1.2.: from the Plymouth strain: England
	1052/6	Droop 1951: marine: Finland

PHILODINA Ehrenberg (rotifer)

P. acuticornis *var.* **odiosa**

LB	R1	originally from Carolina Biological Supply Company

PHORMIDIUM Kützing

P. autumnale (Agardh) Gomont)
LB 1462/6 Lefèvre 1930: France

P. foveolarum Gomont
1462/1 De 1939 (*Schizothrix calcicola var. glomerulata* Baker et Bold)

P. inundatum (Kütz.) Gom.
LB 1462/9 Lefèvre 1952: thermal: used as food for ciliates

P. luridum *var.* **olivaceum** Boresch
1462/2 Boresch. type material: identified as *Plectonema boyranum* by Allen and as *Schizothrix calcicola var. glomerulata* by Baker *et* Bold

P. minnesotense (Tilden) Drouet
LB 1462/3 Myers, Tx 7

P. periscinum (Reinke) Gomont
B 1462/5 Provasoli 1954

P. uncinatum (Agardh) Gomont
LB 1462/7 Lefèvre 1936: France

P. sp.
1462/8 Lewin 1965: marine: California USA

PHYSOLINUM Printz see *Trentepohlia*

PILINIA Kützing

?P. sp. Kützing
461/1 Lewin 1953: marine: Canada

PINNULARIA Ehrenberg

P. maior (Kütz.) Rabenhorst
LB 1056/1 Munch 1975: Maryland USA

PITHOPHORA Wittrock

P. oedogonia (Montagne) Wittrock *var.* **polyspora** Rendle *et* West
LB 530/1 George 1954: Hong Kong

PLAGIOSELMIS Butcher

P. punctata Butcher
LB 990/1 Butcher: Plymouth No. 172A: marine

PLATYAMOEBA Page

P. bursella Page
B 1565/5 Page 1971: type material: marine: England

P. calycinucleolus Page
B 1565/6 Page 1972: type material: marine: England

P. flabellata Page
B 1565/4 Page 1972: type material: marine: England

P. mainensis Page
B 1565/1 Page 1969: type material: marine: Maine USA

P. placida Page
B 1565/2 Page: type material: Wisconsin USA

P. plurinucleolus Page
B 1565/7 Page 1972: type material: marine: England

P. stenopodia Page
B 1565/3 Page: type material: Alabama USA

PLATYDORINA Kofoid

P. caudata Kofoid

LB	160/1a	Wilbois 1957: Indiana No. 850: female: Iowa USA
LB	160/1b	Wilbois 1957: Indiana No. 851: male: Iowa USA
LB	160/1c	Harris 1965, Indiana No. 1659: female: Kansas USA
LB	160/1d	Harris 1965, Indiana No. 1658: male: Kansas USA
LB	160/1e	Harris 1965, Indiana No. 1660: male: Kansas USA
LB	160/1f	Harris 1965, Indiana No. 1661: female: Kansas USA

PLATYMONAS West (see also *Tetraselmis*)

P. convolutae Parke *et* Manton

66/9	Provasoli 1958 *ex Convoluta roscoffensis*, Plymouth No. 372: marine
66/10	Jowett/Cann 1966, Plymouth No. 372a: type material: marine: France

P. gracilis Kylin
161/4 Littler 1973: marine: California USA

P. impellucida McLachlan *et* Parke
161/5 Pintner/Cann 1959, Plymouth No. 429: type material: marine: Puerto Rico

P. subcordiformis (Wille) Hazen

161/1a	Lewin 1952: marine: USA
161/1b	Strout 1952: marine: USA
161/3	from University of California, Irvine, as *Platymonas* A: originally from Caltech

P. spp.

66/3	Wood/Droop 1957: Millport 115: marine: N.S.W. Australia
66/8	Parke 1965: Plymouth No. 315: marine: England
66/11	Jowett 1965: Plymouth No. 362: as *Pyramimonas sp.*: marine: England

PLECTONEMA Thuret *ex* Gomont

P. battersii Gomont
1463/3 Lewin 1965: marine: California USA

P. boryanum Gomont

1446/2 Dyer, B.G. 15, strain 55: H.M.S. Z. 55.3.1
1446/3 Dyer, B.G. 16, strain 6: H.M.S. Z. 55.2.1
1462/2 Boresch as *Phormidium luridum var. olivaceum* Boresch
1462/4 Dyer, strain 76 as *Phormidium*
1463/1 Dyer H.M.S. Z. 59.1.1. as *Plectonema notatum*
1463/2 Lewin: from soil: Novia Scotia Canada

PLEODORINA Shaw

P. californica Shaw

LB 162/1 Starr 1953: homothallic: Tennessee USA

P. illinoisensis Kofoid (*Eudorina illinoisensis* (K.) Pascher)

L 162/2a Stein 1956: male: Minnesota USA
L 162/2b Stein 1956: female: Minnesota USA

PLEURASTRUM Chodat

P. obovatum (Vischer) Tupa

445/1 Vischer 1928, No. 46: Switzerland

P. paucicellulare Vischer

463/1 Vischer 1930, No. 68: Switzerland

P. terrestre Fritsch *et* John

463/2 Pringsheim 1940: material from John: type material: from soil: England

P. terrestre *var.* **indica** Mitra

463/3 Pringsheim: material from Mitra

PLEUROCHLORIS Pascher

P. commutata Pascher

860/1a Vischer 1940, No. 241: from soil: Switzerland
B 860/1b Flint, No. 27: from soil: New Zealand

P. magna Boye-Petersen

B 860/4 Flint, No. 32: from soil: New Zealand

P. meiringensis Vischer

860/3 Vischer 1945, No. 368: from soil: Switzerland

PLEUROCHRYSIS Pringsheim

P. scherffelii Pringsheim (= *Cricosphaera carterae*)

LB 944/1 Pringsheim 1951: type material: brackish: England

PLEUROCOCCUS Meneghini

P. sp.

	464/1	Chodat: Vischer No. 146

PLEUROSIGMA W. Smith *emend.* Cleve

P. sp.

LB	1058/1	Belcher 1974: marine

PLEUROTAENIUM Naegeli

P. minutum (Ralfs) Delponte

LB	664/1	George 1949: Ireland

PODOPHRYA Ehrenberg see *Discophrya*

POLYEDRIELLA Pascher

P. helvetica Vischer *et* Pascher

	861/1	Chodat, Geneva No. 255: Vischer No. 170

POLYEDRIOPSIS Schmidle

P. bitridens (Beck-Mannagetta) Kováčik

	282/1	Starr: soil: USA

POLYTOMA Ehrenberg

P. uvella Ehrenberg

	62/2a	Pringsheim, strain 2
	62/2b	Pringsheim, strain 13
	62/2c	Pringsheim, 1941, strain 1: England
LB	62/2d	Pringsheim
LB	62/2e	Pringsheim 1941: England
LB	62/2f	Pringsheim 1940, strain 4
LB	62/2h	Pringsheim, strain 10
LB	62/2k	Pringsheim 1941, strain M: from manure
	62/2m	Pringsheim: Holland

POLYTOMELLA Aragao

P. caeca Pringsheim

LB	63/2a	Pringsheim pre 1937, material from Kniep: Germany

P. caeca *var.* **minor** Pringsheim

LB	63/2b	Pringsheim 1937: Czechoslovakia

P. magna Pringsheim

	63/3	Pringsheim 1947: from *Ulmus* sap: type material: England

P. papillata Pringsheim

LB 63/2c Pringsheim 1944, material from Sister Monica Taylor: type material: Scotland

P. parva Pringsheim

LB 63/1 Pringsheim: soil: England

PORPHYRA Agardh

P. linearis Greville (*Conchocelis* Stage)

LB 1379/1 Chen 1969: Canada

P. miniata (Agardh) Ag. (*Conchocelis* Stage)

LB 1379/2 Chen 1968: Canada

PORPHYRIDIUM Naegeli

P. aerugineum Geitler

1380/2 Starr: USA

P. purpureum (Bory) Ross (= *P. cruentum*)

1380/1a Vischer 1935, No. 107: Switzerland: halophilic

1380/1b halophilic

1380/1c halophilic

POTERIOOCHROMONAS Scherffel

P. malhamensis (Pringsheim) Peterfi

L 933/1a Chen 1948: type material: England: much used as *Ochromonas malhamensis*

L 933/1b Provasoli

L 933/1c Provasoli

P. nutans Jane

L 933/1d Lewin 1950, formerly listed as *P. stipitata*

P. sociabilis (Pringsheim) Peterfi

L 933/3 Pringsheim: type material

PRASINOCLADUS Kuckuck

P. marinus (Cienk.) Waern

163/1a Parke/George 1954, Plymouth No. 79: marine: England

163/1b Parke/Cann 1963, Plymouth No. 308: marine: England

P. verrucosa (Butcher) Parke

163/3 Butcher/Cann, Plymouth No. 396, sent there by Provasoli: type material: marine

163/4 Butcher/Pennick: marine

P. sp.

163/2 Provasoli/Cann, Ca-5, Plymouth No. 371: marine

PRASIOCOCCUS Vischer

P. calcarius (B. Petersen) Vischer
365/1 Vischer 1950: received from Christensen: Germany

PRASIOLA Agardh

P. stipitata Suhr
468/1 Lewin 1953: marine: Nova Scotia, Canada

PROROCENTRUM Ehrenberg *em.* Dodge

P. balticum (Lohmann) Loeblich)
LB 1114/1 Parke 1950 as *Exuviaella baltica*, Plymouth No. 28: marine: England

P. dentatum Stein
LB 1136/3 Jowett 1965, Plymouth No. 386 as *P. obtusidens*: marine: England

P. micans Ehrenberg
LB 1136/1 Parke 1956, Plymouth No. 97: marine: England
LB 1136/4 Adams 1957, Plymouth No. 97a: marine: England
LB 1136/6 F. Ott, Va 5: marine: Virginia USA

P. minimum (Pavillard) Schiller
LB 1112/1 Parke (as *Exuviaella mariae-lebouriae*) Plymouth No. 18: type material: marine: England
LB 1136/7 F. Ott, Va 13: marine: Virginia USA

P. nanum Schiller
LB 1114/2 Adams 1957 (as *Exuviaella pusilla*) Plymouth No. 184: marine: England

P. triestinum Schiller
LB 1136/5 Jowett 1965, Plymouth No. 394: marine: England

PROTOSIPHON Klebs

P. botryoides (Kützing) Klebs
731/1a Pringsheim, material from Pascher: monoecious
731/1b Cowan 1951: monoecious: Connecticut USA
731/2 Pringsheim, material from Moewus: dioecious, only one strain extant

P. botryoides *f.* **parieticola** Iyengar
731/3 Mitra

PROTOTHECA Krüger. *Some of these strains are probable human pathogens. All should be handled with the greatest care*

P. chlorelloides Beijerinck
N 263/1 ?Beijerinck: received from Delft 1948: ?= *P. zopfii*

P. krugeri

N	263/6	from Delft: isolated from *Xanthoria parietina*

P. moriformis Krüger

N	263/2	? Krüger: received from Delft 1948

P. portoricensis Ciferri, Ashford, and Dalmau

N	263/3a	Ashford: type material
N	263/3b	Parker 1953: from milk: England

P. portoricensis *var.* **trisporus** Ciferri, Ashford, and Dalmau

N	263/4	Ashford: type material

P. zopfii Krüger

N	263/5	Pringsheim: ?= *P. portoricensis*

P. sp.

N	263/7	Hong Kong

PRYMNESIUM Massart

P. parvum Carter

LB	946/1b	Butcher 1952, Plymouth No. 94: marine: toxic to fish: England
	946/1d	Reich 1954: Israel

P. sp.

LB	946/2	Butcher 1958: marine: England

PSEUDANABAENA Lauterborn

P. brunea ? *auct.*

LB	1464/2	Pringsheim 1956

P. catenata Lauterborn

B	1464/1	Pringsheim 1940: England

P. sp.

LB	1464/3	Parke, Plymouth No. 114: marine

PSEUDENDOCLONIOPSIS Vischer

P. botryoides Vischer

	465/1	Vischer 1929, No. 21: type material: Switzerland

PSEUDENDOCLONIUM Wille

P. basiliense *var.* **brandii** Vischer

	466/2	Vischer 1933, No. 52: type material: Switzerland

PSEUDOCHARACIOPSIS Lee *et* Bold

P. texensis Lee *et* Bold

	864/1	Tupa, ? 1969: type material: Texas USA

PSEUDOCHLOROCOCCUM Archibald

P. polymorphum Archibald
265/2 Archibald: type material: soil

P. typicum Archibald
B 265/1 Archibald: type material: soil

PSEUDOCOCCOMYXA Korshikov

P. adhaerens Korshikov
812/2a Simmonds 1962: from potato plant: England
812/2b Simmonds 1964: from tomato plant: England. Several strains listed as *Coccomyxa* probably belong here

PSEUDOHOLOPEDIA Elenkin

P. convoluta (Brébisson) Elenkin
LB 1481/1 Pringsheim, as *Tetrachloris sp.*

PSEUDOISOCHRYSIS *nom. prov.* (F. Ott)

P. paradoxa *nom. prov.* (F. Ott) (resembles *Isochrysis litoralis* Billard *et* Gayral
LB 949/1 Ott, Va 12: marine: Virginia USA

PSEUDOPEDINELLA Carter

P. spp.

LB	947/1a	Jowett 1965, Plymouth No. 361A: marine: England
LB	947/1b	Jowett 1965, Plymouth No. 361: marine: England
LB	947/2	Jowett 1967, Plymouth No. 439: marine: England
LB	947/3	Jowett 1968, Plymouth No. 451: marine: England
LB	947/4a	Butcher/Parke 1959, Plymouth No. 214: marine: England
LB	947/4b	Butcher, received before 1965: marine
LB	947/5	Jowett 1967, Plymouth No. 436: marine: England
LB	947/6	Jowett 1967, Plymouth No. 437: marine: England
LB	947/7	Jowett 1967, Plymouth No. 438: marine: England
LB	947/8	Jowett 1965, Plymouth No. 343: marine: England
LB	947/9	Jowett 1966, Plymouth No. 413: marine: England
LB	948/1	Parke, Plymouth No. 91: marine: England: received as *Pseudopedinella* in 1954: identity now doubtful

PSEUDOPLEUROCOCCUS Snow

P. printzii Vischer
467/1 Vischer 1926, No. 10: type material: Switzerland

PSEUDOSTICHOCOCCUS Moewus

P. monallantoides Moewus
364/1 Moewus 1950: type material: marine: Germany

PSEUDOTREBOUXIA Archibald

P. aggregata Archibald
219/1d — Quispel: *ex Xanthoria parietina*

P. decolorans (Ahmadjian in Ed.) Archibald
219/4 — di Benedetto *ex Xanthoria parietina*: type material of *Trebouxia albulescens*: Italy
219/5a — Ahmadjian *ex Xanthoria parietina*: type material of *Trebouxia decolorans*: University of Indiana No. 901

P. incrustata (Ahmadjian in Ed.) Archibald
219/6 — Ahmadjian *ex Lecanora dispersa*: type material: University of Indiana No. 784

P. potteri (Ahmadjian in Ed.) Archibald
219/7 — Ahmadjian: type material: University of Indiana No. 900

PTEROMONAS Seligo

P. angulosa Lemmermann
LB 64/3 — Pringsheim 1948: England

P. angulosa Lemm. *var.* **takedana** (West) Pascher
LB 64/2 — Pringsheim 1944: England

P. protracta Lemmermann
LB 64/1 — Pringsheim 1942: England

P. varians Jane
LB 64/4 — Evans 1961: England

PTEROSPERMA Pouchet

P. nationalis Lohm.
LB 165/2 — Parke 1963, Plymouth No. 302: marine: England

PYRAMIMONAS Schmarda

P. amylifera Conrad
LB 67/3 — Adams 1961, Plymouth No. 246: marine: England

P. grossii Parke
LB 67/10 — Reynolds 1970, Ci 11: marine: Scotland
LB 67/11 — Pennick 1974: marine: England

P. obovata N. Carter
LB 67/6 — Parke 1962, Plymouth No. 280: marine: England

P. aff. orientalis Butcher

LB 4/1 Butcher, pre 1957 as type material of *Asteromonas propulsum*: marine
LB 67/8 Butcher 1960, as *P. disomata*: marine: England
LB 67/9 Butcher 1956, as *P. disomata*: marine: England
LB 67/12 Butcher: marine: England
LB 67/14 Van der Veer 1967, Plymouth No. 453: marine: England
LB 67/18 Butcher: marine: England
LB 67/19 Butcher: marine: England
LB 67/20 Pennick 1975: marine: England
LB 11/92 Butcher 1959 as ? *Chlamydomonas*: marine: England

P. parkeae Norris *et* Pearson

LB 67/7 Parke 1963, Plymouth No. 299: material from Norris: marine: USA
LB 67/17 Pennick 1974, *ex* Plymouth No. 340: marine: England
LB 67/15 Plymouth "PP": marine
LB 67/21 Norris 1966, University of Washington 15.8.25, Plymouth No. 492: type material: marine: California USA
LB 135/2 Adams 1958, Plymouth No. 205: as *Halosphaera minor*: marine: England

P. tetrarhynchus Schmarda

LB 67/4 Belcher and Swale 1971: England

P. virginica *nom. prov.*

LB 67/16 Ott: marine: Virginia USA

PYROBOTRYS Arnoldi

P. stellata Korshikov

LB 10/1c Pringsheim 1956: Florida USA

QUADRIGULA Printz

Q. closterioides (Bohlin) Printz

268/1 George 1949, material from Pringsheim: Ireland

RADIOSPHAERA Snow

R. dissecta (Korshikov) Starr

B 247/1 Starr 1951: from soil: England
3/2b Vischer 1942, No. 321: from soil, as *Asterococcus (Macrochloris) terrestris* Vischer

R. negevensis Ocampo-Paus *et* Friedmann *f.* **negevensis**

247/2 Ocampo-Paus: type strain 1-110: soil: Negev Desert

R. negevensis *f.* **minor** Ocampo-Paus *et* Friedmann

247/3 Ocampo-Paus: type strain 1-117: soil: Negev Desert

R. sphaerica (Korshikov) Fott

3/2b Vischer 1942, No. 321 (according to Fott 1971)
247/1 Starr 1951 (according to Fott 1971)

RAPHIDIOPHRYS Archer

R. ambigua Penard
LB 1568/1 Page 1974: England

RAPHIDONEMA Lagerheim

R. longiseta Vischer
470/1a Vischer 1932, No. 97: type material: Switzerland
470/1b Pringsheim

R. spiculiforme Vischer
470/2a Vischer 1940, as No. 208: type material: Switzerland
470/2b George 1951: England
470/2c George 1951: England

R. sp.
470/3 Belcher 1961: England

RHABDOMONAS Fresenius

R. costata (Korshikov) Pringsheim (= *Menoidium longum* Pringsheim)
LB 1271/1 Pringsheim: Czechoslovakia

R. gibba (Skuja) Pringsheim
LB 1271/2 Pringsheim 1940: England

R. incurva Fresenius *var.* **major** Pringsheim
LB 1271/4 Pringsheim: Czechoslovakia

R. spiralis Pringsheim (= *Rhabdospira*)
LB 1271/5 Pringsheim 1936: Austria

RHABDOSPIRA Pringsheim

R. spiralis Pringsheim
LB 1271/5 Pringsheim 1936: type material: Austria

RHIZOCLONIUM Kützing

R. hieroglyphicum (Agardh) Kützing
LB 540/1 George 1950: from soil: England

RHODELLA Evans

R. maculata Evans
LB 1388/2 Droop, Millport 207: type material: marine

RHOPALOCYSTIS Schussnig

R. sp.
274/1 Flint 1962/King: soil: New Zealand

RHYNCHOMONAS Klebs

R. nasuta (Stokes) Klebs
B 1972/1 Page 1972: England

ROSCULUS Hawes

R. ithacus Hawes
B 1571/1 Fennell 1974: from *Zea mays*: Illinois USA

SACCAMOEBA Frenzel

S. limax (Dujardin)
B 1534/6 Page 1967, No. 57: Alabama USA

S. stagnicola Page
B 1572/1 Page 1972: type material: England

SARCINOCHRYSIS Geitler

S. granifera (Mack) Tschermak-Woess
LB 952/1 Tschermak-Woess 1970: saline: Austria

SCENEDESMUS Meyen

S. acuminatus (Lagerheim) Chodat
276/12 Algéus ?1950: Sweden

S. acutiformis Schroeder
276/11 Algéus ?1950: ?Sweden

S. acutus *f.* **alternans** Hortobagyi
276/3a Pringsheim, as *S. obliquus*

S. basiliensis Chodat
276/1a Chodat: received from Vischer
276/1b Chodat: received from Vischer

S. bijugatus (Turpin) Kützing *var.* **seriatus** Chodat
276/14 L. Moewus

S. dimorphus Kützing
276/10 Algéus 1950: Sweden

S. dispar Brébisson
276/13 George 1950: material from Pringsheim: Sweden

S. naegelii Chodat
276/2 Rodhe No. 1598.1: Sweden

S. obliquus (Turpin) Krüger
276/3b Algéus 1939: ?Sweden
276/3c Pirson, as M-B

S. opoliensis P. Richter

	276/15	Atkinson/Baker 1960: Malacca

S. pannonicus Hortobagyi

	276/4a	Pringsheim as *S. quadricauda*: Czechoslovakia

S. quadricauda (Turpin) Brébisson

	276/4b	Pringsheim 1940: England
	276/4c	Rodhe No. 1621a: ?Sweden
	276/4d	Algéus 1950: ?Sweden
	276/4e	Pirson as M-A

S. spp.

	276/5	Pringsheim
	276/6a	Gaffron, clone D3
	276/6b	Gaffron, clone D3: received from A.H. Brown 1957
	276/7	Pringsheim 1940: England
	276/8	Lewin 1950: Connecticut USA
	276/9	Harder: ?Germany
	211/23	Lewin, Indiana No. 343: type of *Chlorella fusca*: shown to be a unicellular *Scenedesmus* by Fott *et al* (1975)

SCHIZOMERIS Kützing

S. ? leibleinii Kützing

	376/1	Christensen No. 9131/Baker: Kentucky, USA

SCHIZOTHRIX Kützing *ex* Gomont

S. calcicola (Ag.) Gomont

	1470/1	Myers, Tx 27: identified by Drouet
B	1470/3	Pentecost 1972: tufa deposit: England

S. calcicola *var.* **glomerulata** Baker *et* Bold

	1462/1	De 1939: identified by Baker and Bold
	1462/2	Boresch

S. sp.

LB	1470/2	Ott 0404: USA

SCOTIELLA Fritsch

S. oocystiformis Lund

	277/1	George 1955: material from Lund: England

SCYTONEMA Agardh *ex* Bornet *et* Flahault

S. javanicum (Kützing) Bornet

	1473/1	George 1964: from material epiphytic on *Peperomia* leaf

S. ocellatum Lyngbye

B	1473/2	Ott/Baker: Ott 317: soil: Hawaii

S. spp.

B	1473/3	George 1968: soil: Japan
LB	1410/4	*Calothrix anomala*, may belong here

SELENASTRUM Reinsch

S. bibraianum Reinsch

	278/1	Pringsheim 1940: England

S. capricornutum Printz

	278/4	University of Indiana No. 1648: ?Norway

S. gracile Reinsch

	278/2a	George 1947: England
	278/2b	George 1954: Nigeria
	278/2c	Starr: Tennessee USA
	278/2d	Cowan: Connecticut USA
	278/2e	Bourrelly, Paris No. 84
	278/2f	Wurtz, No. 16: France
	278/2g	Lewin, Connecticut USA
	278/2h	Starr: Tennessee USA

S. minutum (Naegeli) Collins

	278/3	Myers, No. Tx 1: Texas, USA

SKELETONEMA Greville

S. costatum (Greville) Cleve

LB	1077/1a	Droop: Millport 73
LB	1077/1b	Robinson 1950, Plymouth No. 106: marine: England

SPERMATOZOPSIS Korshikov

S. exsultans Korshikov

LB	175/1	Belcher and Swale: England

SPHAERELLA Sommerfeldt = *Haematococcus* Flotow

SPHAERELLOPSIS Korshikov = *Vitreochlamys* Batko

SPHAEROPLEA Agardh

S. soleirolii (Duby) Montagne

LB	377/1a	Starr 1951: England

S. wilmanae Fritsch *et* Rich

LB	377/1b	George 1951: South Africa
LB	377/1c	George 1951: South Africa
LB	377/1d	George 1951: South Africa
LB	377/1e	George 1951: South Africa

SPHAEROSORUS Pascher

S. composita Moewus

	876/1	Moewus: marine: type material

SPIRILLUM Ehrenberg (bacteria)

S. serpens (Müller) Winter
LB 1780/1 Pringsheim

SPIROGYRA Link

S. grevilleana (Hassall) Kützing
LB 678/1 George 1947: England

S. majuscula Kützing
LB 678/2 George 1947: England

S. pratensis Transeau
LB 678/7a Allen No. 10: diploid: Indiana No. 924: Indiana USA
LB 678/7b Allen No. 100: tetraploid: Indiana No. 926: Indiana USA
LB 678/7c Allen No. 103: tetraploid: Indiana No. 927: Indiana USA

S. varians (Hassall) Kützing
LB 678/3 George 1947: England

S. spp.
LB 678/4 Pringsheim: resembles *S. grevilleana* but has plain end walls: Czechoslovakia
LB 678/6 George 1956: an attached species: Ireland
LB 678/8 Peck 1967: England

SPIROSTOMUM Ehrenberg

S. ambiguum Ehrenberg
LB 1677/2b George 1955: England

S. intermedium Kahl
LB 1677/3 George 1951: England

S. teres Claparède *et* Lachmann
LB 1677/1 Chen 1949: England

SPIRULINA Gomont

S. geitleri de Toni
LB 1475/4b Wood: alkali lake: Abyssinia
LB 1475/4c Clement, T274: Lake Chad

S. major Kützing
LB 1475/3 George 1953: brackish: England

S. platensis (Nordstedt) Geitler (*Arthrospira platensis*)
LB 1475/4a George 1955: alkali lake: Mexico

S. subsalsa Oerstedt *ex* Gom. *f.* **versicolor**
LB 1475/2 Pringsheim 1949: marine: England

S. sp.
LB 1475/1 Pringsheim 1951: *aff. S. major*: marine: England

SPONDYLOSIUM Brébisson

S. pulchellum Archer
LB 680/1 George 1952: Ireland

SPONGIOCHLORIS Starr

S. excentrica Starr
280/1 Bold: type material; soil; Tennessee USA

S. spongiosa (Vischer) Starr
3/1 Vischer 1942, No. 318: type material: soil: Switzerland
3/2a Pringsheim: material from John as *Asterococcus terrestris*

SPUMELLA Cienkowski

S. elongata (Stokes) Belcher *et* Swale
LB 955/1 Belcher and Swale 1974: soil: England

STACHYAMOEBA Page

S. lipophora Page
B 1579/1 Darbyshire 1973: type material; soil: Scotland

STAURASTRUM Ralfs

S. gracile Ralfs
LB 679/3 Rodhe: Sweden

S. ? muricatum Brébisson
LB 679/5 George 1956: Ireland

S. orbiculare Ralfs ? *var.* **ralfsii** W. *et* G.S. West
679/2 Ondraček, as Cosmarium No. 4: Prague No. 230

S. punctulatum Brébisson
679/1 Czurda 1930: Prague No. 226

S. sp. *aff. S. teliferum*
LB 679/4 King 1953: Wales

STENTOR Oken

S. coeruleus Ehrenberg
LB 1682/1 George 1963: England

S. polymorphus (Müller)
LB 1682/2 Goodfellow 1975: England

STEPHANODISCUS Ehrenberg

S. sp.
LB 1079/1 Belcher 1974: England

STEPHANOSPHAERA Cohn

S. pluvialis Cohn

LB	78/1a	Pringsheim 1950: Sweden
LB	78/1b	Pringsheim 1950: Sweden
LB	78/1c	Pringsheim 1950: Sweden
LB	78/1d	Pringsheim 1950: Sweden

STICHOCOCCUS Naegeli

S. bacillaris Naegeli

	379/1a	Vischer
	379/1b	Vischer 1923, No. 1: Switzerland
	379/1c	Algéus: Sweden
	379/1d	Lewin 1952: Alaska
	379/1e	Gray 1956: from freeze-dried contents of a cow's rumen: England
	379/5	Parke 1949/George 1954, Plymouth No. 82: marine: type of material of *S. cylindricus* Butcher: England

S. chloranthus Krüger

	379/2	Krüger: type material

S. cylindricus Butcher

	379/5	as *S. bacillaris* q.v.

S. ? fragilis Gray

	379/4	Lewin 1951: Massachusetts USA

S. mirabilis Largerheim

	379/3	Pringsheim

STIGEOCLONIUM Kützing

S. ? amoenum Kützing

	477/8	Reynolds No. 6

S. farctum Berthold

LB	477/10a	Reynolds No. 1: Wales
LB	477/10b	Butcher: Reynolds No. 15b

S. helveticum Vischer *var.* **minus** Vischer (? = *S. amoenum*)

	477/1	Vischer 1925 No. 24: type material: Switzerland

S. helveticum *var.* **maius** Vischer (? = *S. amoenum*)

B	477/2	Vischer 1926 No. 26: type material: Switzerland

S. ? huberi Heering

	477/7	Reynolds No. 2

S. nanum Kützing

LB	477/18	Butcher: Reynolds No. 8b: England

S. pascheri (Vischer) Cox *et* Bold

	410/1	Vischer 1928, No. 45: type material of *Caespitella pascheri*: Switzerland
	410/2	Lewin 1950: Connecticut USA

S. tenue Kützing

LB	477/11a	Reynolds, No. 4: Wales
LB	477/11b	Butcher: Reynolds No. 1b: England

S. ? variabile Naegeli

LB	477/13	Butcher: Reynolds No. 9b: England

S. spp.

	477/3	Lewin, No. 4: resembles *S. helveticum*: USA
	477/4	Lewin, No. 3: USA
	477/6	Lewin, No. 6: USA
	477/9	Reynolds, No. 7: resembles *S. farctum*
LB	477/12	Reynolds, No. 4: a very slender form: ? *S. setigerum*: Wales
LB	477/14	Reynolds 1950, No. 28: a very long erect system: England
LB	477/15	Reynolds, No. 15: with conspicuous hairs: Wales
LB	477/16	Reynolds, No. 25: prostrate system grows indefinitely: Wales
LB	477/17	Reynolds, No. 16: a large plant, approaching *Draparnaldia* in morphology: Wales
	477/19a	Lewin, No. 102: Massachusetts USA
	477/19b	Lewin, No. 104: Massachusetts USA
	477/19c	Lewin, No. 106: Massachusetts USA
	477/19d	Lewin, No. 109: Massachusetts USA
	477/19e	Lewin, No. 112: Massachusetts USA
	477/20	Vischer, No. 139: ? Switzerland
	477/21	Vischer, No. 115: ? Switzerland
	477/22	Vischer, No. 579: epiphyte from *Cladophora*: ? Switzerland
	477/23	Vischer, No. 195: ? Switzerland

STREBLONEMA Derbès *et* Solier

S. sp.

	1337/1	Lewin: marine: California USA

STROMBOMONAS Deflandre

S. conspersa (Pascher) Deflandre

LB	1280/1	Pringsheim 1943: England

STYLONYCHIA Ehrenberg

S. mytilus (Müller) (syngen 1)

LB	1681/2	Ammermann 1975: Germany

SYMPLOCA Kützing

S. muscorum (Agardh) Gomont

B	1478/1	George 1955: British Columbia Canada

SYNCRYPTA Ehrenberg *emend.* Bourrelly

S. glomerifera Clarke *et* Pennick

LB	958/1	Butcher, FH14: type material: marine: England

SYNECHOCOCCUS Naegeli

S. cedrorum Sauv.

B 1479/2a Gif strain M 137/1a

S. elongatus Naegeli (= *Anacystis marina* Drouet *et* Daily)

LB 1479/1a Pringsheim 1940
1479/1b Gif M 137/2

S. leopoliensis (Racib.) Komárek

1405/1 Kratz/Allen: see Komárek 1970 for an account of the generic identity of this strain which has been much used under the name "*Anacystis nidulans*"

S, sp.

1479/4 Jaworski 1974: L270: England

SYNECHOCYSTIS Sauvageau

S. minima Bourr.

B 1480/1 Gif No. A.M. 138/2

SYNURA Ehrenberg

S. petersenii Korsh.

LB 960/1a Pringsheim strain 1
LB 960/1b Pringsheim strain 2
LB 960/1c Pringsheim strain 6

S. uvella Ehrenberg

LB 960/2 Pringsheim strain 4

SYRACOSPHAERA Lohmann

S. carterae Braarud *et* Fagerland = *Cricosphaera carterae* **q.v.**

TABELLARIA Ehrenberg

T. flocculosa (Roth) Kützing *var.* **asterionelloides** (Grun. *in* V.H.) Knudsen

LB 1081/1 Jaworski 1974, FBA L228: England

TETRACYSTIS Brown *et* Bold

T. aeria Brown *et* Bold

181/1a Brown 1960: type material: from air over Texas USA
181/1b Brown 1960: from air over Texas USA

T. aggregata Brown *et* Bold

181/2 Brown 1960: type material: from air over Texas USA

T. aplanosporum (Arce *et* Bold) Brown *et* Bold

181/9 Arce: type material: Indiana No. 773: from soil: Cuba

T. dissociata Brown *et* Bold

207/1b Vischer No. 304 as *Borodinella*: type material: Indiana No. 128: via Brown

T. excentrica Brown *et* Bold

181/3 Brown 1961: type material: from soil: Colorado USA

T. illinoisensis Brown *et* Bold

181/4 Brown 1962: type material: from air over Illinois USA

T. intermedium (Deason *et* Bold) Brown *et* Bold

181/10 type material: Indiana No. 974

T. isobilateralis Brown *et* Bold

181/5 Johnston 1960: type material: from soil: Texas USA

T. pampae Brown *et* Bold

181/6 Brown 1961: type material: from soil: Texas USA

T. pulchra Brown *et* Bold

181/7 Sweet 1962: type material: from soil: Texas USA

T. tetrasporum (Arce *et* Bold) Brown *et* Bold

181/11 Arce: type material: Indiana No. 780: from soil: Cuba

T. texensis Brown *et* Bold

181/8 Mattox: type material: from soil: Texas USA

TETRAEDRON Kützing

T. bitridens Beck Mannagetta = *Polyedriopsis bitridens* (Beck Mannagetta) Kováčik

282/1 Starr 1952: University of Indiana No. 120: Massachusetts USA

TETRAHYMENA Furgason. Some of these strains were formerly known as *Glaucoma*

T. patula (Müller) Corliss

L 1630/2 L-FF Fauré Fremiet/Lwoff 1942: selected as type material by Corliss (1971): France

T. pyriformis (Ehrenberg) Lwoff

L	1630/1	Ch-S, Chatton 1925: France
L	1630/1	E, Elliott 1932: selected as type material by Corliss (1971): New York USA
L	1630/1	Gf-J, Johnson 1934: New York USA
L	1630/1	GL, Lwoff 1922: France
L	1630/1	G1-R, Robertson 1935: England
L	1630/1	GP, Hetherington 1934: Massachusetts USA
L	1630/1	H, Hetherington 1931: California USA
L	1630/1	HS, Phelps 1949: thermal strain: Texas USA
L	1630/1	L-I, Loefer 1948: Texas USA
L	1630/1	L-II, Loefer 1948: Texas USA
L	1630/1	N, Phelps 1948/9 mating type 1/I, Arizona USA
L	1630/1	N, Phelps 1948/9 mating type 1/II
L	1630/1	N, Phelps 1948/9 mating type 1/III
L	1630/1	N, Phelps 1948/9 mating type 1/IV
L	1630/3	PP, Kidder 1942: Tennessee USA
L	1630/1	S, Seaman 1946: USA
L	1630/1	T, Thomas 1931: California USA
L	1630/3	V_1, Lilly 1940: Rhode Island USA
L	1630/1	W, Claff 1939: Massachusetts USA
L	1630/1	WH-6, Elliott: Massachusetts USA
L	1630/1	WH-14, Elliott: Massachusetts USA

T. vorax (Kidder, Lilly *et* Claff) Kidder

L	1630/3	V_2 Lilly 1947: (polymorphic strain): selected as type material by Corliss (1971): Rhode Island USA

The following *Tetrahymena pyriformis* strains have been characterized electrophoretically by Borden, Whitt and Nanney (1973). The first letter indicates the phenoset as designated by Borden *et al*; the second letter(s) the original strain designation and finally the source is indicated

A/GL/Zeuthen
A/GL/Frankel
A/H/CCAP
B/GL/Eichel
B/E/CCAP
C/GL/ATCC 30006
D/HS/Eichel
D/HSM/Cameron

Other CCAP strains were placed in phenosets as follows:

H S, T and W as phenoset A
E and G1 as phenoset B
ChS, Gf-J, GP and PP as phenoset E
HS (micronucleate) as phenoset G
G1-R as phenoset H

TETRAMITUS Perty

T. rostratus Perty

B	1581/1	Department of Public Health, Austin: Texas USA

TETRASELMIS Stein (see also *Platymonas*)

T. apiculata (Butcher) Butcher

66/15	Butcher/Belcher and Cann: brackish: England
66/17	Butcher/Belcher and Cann: type material: marine
66/20	Butcher/Cann: marine: England

T. carteriiformis Butcher

66/2	Droop 1952/Belcher and Cann, Millport No. 10: type material: supralittoral: Scotland

T. chuii Butcher

8/6	Chu: type material: marine: Scotland
66/21a	Butcher 1958/Belcher and Cann: marine: England
66/21b	Butcher 1960/Belcher and Cann: marine: England
66/21c	Butcher/Belcher and Cann: marine: England

T. gracilis (Kylin) Butcher

66/13	Butcher 1959/Belcher and Cann: marine: England

T. hazenii Butcher (? = *T. tetrathele*)

66/7	Butcher(?)/Belcher and Cann: Plymouth No. 457, sent there by Provasoli 1968

T. inconspicua Butcher

66/19a	Butcher 1955/Belcher and Cann: marine: England
66/19b	Butcher/Belcher and Cann: marine: England
66/19c	Butcher 1959/Belcher and Cann: brackish: England
66/19d	Butcher/Belcher and Cann: marine: England

T. levis Butcher

66/12	Butcher 1956/Belcher and Cann: marine: England

T. rubens Butcher

66/6	Butcher/Belcher and Cann: Plymouth No. 456, sent there by Provasoli 1968: type material: brackish: England
66/18a	Butcher/Belcher and Cann: marine: England
66/18b	Butcher/Belcher and Cann: marine: England

T. striata Butcher

66/16	Butcher/Belcher and Cann: marine: Wales
66/5	Knight-Jones 1946/Belcher and Cann: Plymouth No. 443: type material: marine: Wales

T. suecica (Kylin) Butcher

66/4	Bernhard pre 1962/Parke 1964/Belcher and Cann: Plymouth No. 305: marine: Italy. Recommended as a food organisms
66/22a	Butcher 1959/Belcher and Cann: marine: England
66/22b	Butcher 1960/Belcher and Cann: marine: Guernsey
66/22c	Butcher 1958/Belcher and Cann: marine: England
66/22d	Butcher 1959/Belcher and Cann: marine: England

T. tetrathele (West) Butcher

66/1a	George 1950: brackish: England
66/1b	George 1950: brackish: England
66/1c	Butcher pre 1961/Cann: Plymouth No. 272: marine
66/1d	Butcher 1956/Belcher and Cann: marine: England
66/14	Butcher 1958/Belcher and Cann: marine: England
161/2	Droop 1952: Millport No. 19: brackish: Wales

T. verrucosa Butcher

66/23	Butcher/Belcher and Cann: marine: Wales

T. spp.

	66/24	Butcher/Cann: marine: Adriatic Sea
	66/25	Butcher/Cann: marine: France
	66/26	Butcher/Belcher and Cann: marine: France
	66/27	Butcher/Belcher and Cann: marine: Malta
	66/28	Butcher/Belcher and Cann: marine: Adriatic Sea
	66/29	Butcher/Belcher and Cann: marine: Monaco

TETRASTRUM Chodat

T. staurogeneiformis (Schroeder) Lemm.
284/1 Belcher and Swale 1975: England

THALASSIOSIRA Cleve

T. sp.
LB 1085/1* Belcher 1975: marine: England

THALASSOMONAS Butcher *nom. confus.* = *Micromonas* p.p. (see Parke *et* Rayns 1964)
= *Mantoniella* p.p. (see Desikachary 1972)

THECAMOEBA Fromentel

T. granifera (Greeff)
B 1583/5 Page 1973 (153): soil: neotype: England

T. orbis Schaeffer
B 1583/2 Page 1969 (78) marine: Maine USA

T. proteoides Page
LB 1583/6 Hülsmann/Page 1974: type material: Germany

T. quadrilineata (Carter)
B 1583/7 Page 1971 (113) neotype: England

T. similis (Greeff)
B 1583/8 Page 1974 (180): soil: neotype: England: *multiplies rapidly*

T. sphaeronucleolus (Greeff)
B 1583/3 Page 1967: Alabama USA

T. striata Penhard
B 1583/4 Page 1968: Wisconsin USA

T. terricola (Greeff)
B 1583/9 Page 1974 (173): soil: neotype: England

TOLYPOTHRIX Kützing *ex* Bornet *et* Flahault

T. distorta Kützing
LB 1482/5 George 1962: soil: California USA

T. distorta *var.* **symplocoides** Hansgirg
1482/2 Manten 1948: Utrecht No. P.39: soil: Holland

T. tenuis Kützing

	1482/3a	Watanabe, W1, Nagao Institute No. A.005: N fixing: Japan
B	1482/3b	Watanabe, M29: N fixing: Japan

TRACHELOMONAS Ehrenberg

T. bernardinensis Vischer

LB	1283/5a	Pringsheim, strain 33

T. bulla Stein

LB	1283/6	Pringsheim 1951, strain 39: France

T. deflandrei Pringsheim

LB	1283/7a	Pringsheim, strain 35: type material

T. grandis Singh

LB	1283/20	Singh: type material: Tennessee USA

T. hispida (Perty) Stein *em.* Deflandre

LB	1283/8	Pringsheim, strain 3

T. hispida *var.* **acuminata** Deflandre

LB	1283/9	Pringsheim 1940, strain 4: England

T. hispida *var.* **coronata** Lemmermann

LB	1283/2	Pringsheim 1940, strain 5: England

T. lefevrei Deflandre

LB	1283/10a	Pringsheim 1943, strain 20: England
LB	1283/10c	Pringsheim 1952, strain 42: England

T. oblonga Lemmermann

LB	1283/11	Pringsheim, strain 9

T. oblonga *var.* **punctata** Pringsheim

LB	1283/12	Pringsheim, strain 22: type material

T. pertyi Pringsheim

LB	1283/13	Pringsheim 1945, strain 30: England

T. volvocina Ehrenberg

LB	1283/4b	Pringsheim 1945, strain 28: England

T. volvocinopsis *var.* **spiralis** Pringsheim

LB	1283/17	Pringsheim 1943, strain 21: type material: England

T. zorensis Deflandre

LB	1283/18	Pringsheim 1940, strain 7: England

TREBOUXIA de Puymaly (see also *Pseudotrebouxia* Archibald)

T. albulescens de Nicola *et* di Benedetto (= *Pseudotrebouxia decolorans* (Ahmad.) Arch.
219/4 di Benedetto *ex Xanthoria parietina*: type material: Italy

T. anticipata (Ahmadjian in Ed.) Archibald
219/3 Ahmadjian *ex Parmelia rudecta*: type material: University of Indiana No. 903

T. arboricola de Puymaly
219/1a isolated?: from Delft: Meyer No. 2

T. crenulata Archibald
219/1b Quispel: *ex Parmelia acetabulum*
219/2 Richardson, 927: *ex Xanthoria aureola*: England

T. flava Archibald
219/1c Quispel: *ex Physica pulverulenta*

T. italiana Archibald
219/5b Tomasalli *ex Xanthoria parietina*: Italy

T. spp.
213/1b Beijerinck: Meyer No. 8: formerly listed as *Chlorococcum humicolum*
213/3 Beijerinck: as *Chlorococcum humicolum*: *ex Cladonia sp.*

TRENTEPOHLIA Martius

T. aurea Martius
B 483/1 George 1949: subaerial: Wales

T. monile de Wildeman (*Physolinum*)
460/1 Flint/George 1955: subaerial: Nigeria

T. ? dialepta (Nylander) Heriot
483/2 George 1962: material from Leakey: subaerial: Uganda

T. sp.
B 483/3 George 1964: subaerial: Ireland

TRIBONEMA Derbès *et* Solier

T. aequale Pascher
880/1 Pringsheim: identified by Pascher: Czechoslovakia

T. viride Pascher
LB 880/3 George 1948: England

T. sp.
B 880/2 Christensen 1949: England

TRINEMA Dujardin

T. lineare Penard

LB	1584/1	Hedley 1970: England

ULOTHRIX Kützing

U. crenulata (Kützing) Kützing

LB	335/6	Pringsheim: material from John: England

U. confervicola (Lagerh.) Mattox *et* Bold

	386/2	Pringsheim: Wisconsin USA

U. fimbriata Bold

	384/2	Bold 1955: type material: Tennessee USA

U. gigas (Vischer) Mattox *et* Bold

	386/3	Vischer 1930, No. 69: type material of *Uronema gigas*: Switzerland

U. minuta Mattox *et* Bold

	386/1	Pringsheim: type material: Wisconsin USA

U. subtilissima Rabenhorst, see *Klebsormidium*

U. sp.

	386/4	Pringsheim 1949: type material of *Uronema terrestre* Mitra

URONEMA Lagerheim (alga) = *Ulothrix* Kützing *emend* Mattox *et* Bold

URONEMA Dujardin (ciliate)

U. ? marinum

LB	1686/2	Burkhill 1973: marine: England

U. schewiakoffi Buddenbrock

LB	1686/1	Parke, Plymouth No. 118: marine: England

UVA Playfair = *Pyrobotrys* Arnoldi

VACUOLARIA

V. virescens Cienkowski

LB	1195/1	Pringsheim 1950: England

VAHLKAMPFIA Chatton *et* Lalung-Bonnaire

V. aberdonica Page

B	1588/4	Darbyshire 1972: type material: soil: Scotland

V. avara Page

B	1588/1a	Page 1964: type material: Indiana USA
B	1588/1b	Page 1964: type material: Washington USA

V. enterica Page

B	1588/5	Kadlec 1972: type material: from **intestine** of a **turkey**: Czechoslovakia

V. inornata Page

B	1588/2	Page 1964: type material: Wisconsin USA

V. jugosa Page

B	1588/3a	Page 1964: type material: Idaho USA
B	1588/3b	Darbyshire 1974: soil: Orkney Islands
B	1588/3c	Darbyshire 1974: soil: Scotland
B	1588/3d	Darbyshire 1974: soil: Scotland
B	1588/3f	Darbyshire 1974: soil: Orkney Islands

V. ustiana Page

B	1588/6	Černa 1972: type material: Czechoslovakia

VANNELLA Bovee

V. mira (Schaeffer)

B	1589/1	Page 1964: Wisconsin USA

V. platypodia (Gläser)

B	1589/2	Page 1964: Indiana USA

V. simplex (Wohlfarth-Bottermann)

B	1589/3	Hülsmann: Germany

VAUCHERIA de Candolle

V. debaryana Woronin

LB	745/5	Christensen 1949: England

V. geminata de Candolle

LB	745/4	Christensen

V. sessilis (Vaucher) de Candolle

LB	745/1b	Christensen
LB	745/1c	Pringsheim

V. woroniniana Heering

LB	745/3	Christensen 1949

VEXILLIFERA Schaeffer

V. bacillipedes Page

B	1590/1	Page 1968: type material: Wisconsin USA

VISCHERIA Pascher

V. punctata Vischer

	887/1	Vischer 1941, No. 266: soil: Switzerland

V. stellata (Poulton) Pascher

	887/2b	Chodat, No. 185: Vischer No. 169: University of Indiana No. 312: Switzerland

VITREOCHLAMYS Batko

V. incisa (Pringsheim) Batko = *Sphaerellopsis incisa* (P.) Ettl
11/10 Pringsheim 1929: type material of *Chlamydomonas incisa*
Pringsheim: Czechoslovakia

VOLVOX Linnaeus

V. aureus Ehrenberg
LB 88/1b George 1960: England

V. globator (L.) Ehrenberg
LB 88/2 George 1949: England

V. tertius Meyer
LB 88/3 George 1947: England

VOLVULINA Playfair

V. steinii Playfair
LB 90/1 Christen 1950: Switzerland

VORTICELLA (L.) Ehrenberg

V. similis Stokes
LB 1690/2 George 1967: England

V. microstoma Ehrenberg
LB 1690/3 George 1975: readily encysts: England

WOLOSZYNSKIA Thompson

W. coronata (Wolos.) Thompson
LB 1117/2 Pringsheim as 'Peridin 3'

W. micra Leadbeater *et* Dodge
LB 1166/1 Adams 1958, Plymouth No. 207: type material: marine: England

ZYGNEMA Agardh

Z. circumcarinatum Czurda
698/1a Czurda 1929: Prague No. 208: + strain: Czechoslovakia
698/1b Czurda 1929: Prague No. 209: - strain: Czechoslovakia

Z. cylindricum Transeau
698/2 Czurda 1929: Prague No. 211: Czechoslovakia

Z. peliosporum Wittrock
698/3a Czurda 1926: Prague No. 206: male strain: Czechoslovakia
698/3b Czurda 1926: Prague No. 205: female strain: Czechoslovakia

Z. sp.

698/4 Pringsheim 1948: England

All recent attempts to obtain conjugation in either of the above pairs have failed

ZYGNEMOPSIS Skuja *em.* Transeau

Z. sp.

LB 699/1 Ott, No. 0379: USA

BRYOPHYTA AND HIGHER PLANTS

AULACOMNIUM Schwaeg

A. androgynum (L.) Schwaeg
1802/1 Pringsheim; ? Czechoslovakia

BRYUM Dill.

B. sp.
1804/1 George 1955: Hong Kong

BUXBAUMIA Haller

B. aphylla Hedw.
1805/1 Keil 1947: Czechoslovakia

B. indusiata Brid.
1805/2 Keil 1948: Czechoslovakia

CERATODON Brid.

C. purpureus Brid.
1807/1 Pringsheim: Czechoslovakia

FOSSOMBRONIA Raddi

F. cristula Austin
1862/2 Morris 1958: New Jersey USA

F. pusilla (L.) Dum.
1862/1 Pringsheim 1939: England

FUNARIA Schreb.

F. hygrometrica Hedw.
1813/1 Pringsheim: Czechoslovakia

HAPLODON R. Br.

H. sp.
1817/1 von Stosch 1938: Germany

HAPLOMITRIUM Nees

H. hookeri (Sm.) Nees
B 1868/1 Lorbeer: female strain: Czechoslovakia. *Grows slowly.*

HYPNUM L.

H. riparium L.
1820/1 Lewin 1954: Nova Scotia

LEPTOBRYUM Wils.

L. pyriforme (Hedw.) Schpr.
1822/1a Pringsheim: ? Czechoslovakia
1822/1b Pringsheim: ? Czechoslovakia
1822/2 Pringsheim: bivalent strain from Wettstein
1822/3 Pringsheim: univalent strain

LOPHOCOLEA Dum.

L. heterophylloides Nees
B 1874/1 Berrie: Australia

MONOSELENIUM Griff.

M. tenerum Griff.
B 1875/1 Lorbeer: Czechoslovakia

PHASCUM Schreb.

P. cuspidatum Schreb.
1831/1a Hughes 1954
1831/1b Whitehouse 1961: bivalent: England

PHYSCOMITRELLA B.& S.

P. patens (Hedw.) B, S, and G.
1833/1 Whitehouse 1961 as PH1: England

RICCIA L.

R. glauca L.
1885/2 Whitehouse 1958: England

R. rhenana Lorbeer *ex* K. Müller
B 1885/1 George 1959: England

R. sorocarpa Bisch.
1885/3 Whitehouse 1960: England

R. duplex
1885/4 G.K.Berrie: Australia

RIELLA Mont.

R. americana Howe *et* Underwood
LB 1887/1a Proctor: male strain: New Mexico USA
LB 1887/1b Proctor: female strain: New Mexico USA

SCHISTOSTEGA Mohr.

S. osmundacea Mohr.

1837/1 Keil. *Grows slowly.*

SPHAEROCARPOS Ludwig

S. texanum Aust.

1890/2 Lorbeer: male strain. *Grows slowly.*

SPHAGNUM Dill

S. squarrosum Pers.

1838/1 Keil 1947: Czechoslovakia

S. sp.

1838/2 Lhotsky 1946: Czechoslovakia

SPLACHNUM L.

S. ampullaceum Hedw.

1839/1a Keil 1946: male strain: Czechoslovakia
1839/1b Keil 1946: female strain: Czechoslovakia

S. sphaericum L.

1839/2a Keil 1946: male strain: Czechoslovakia
1839/2b Keil 1946: female strain: Czechoslovakia

TETRAPHIS Hedw.

T. pelludica Hedw.

1842/1 Lewin 1954: Nova Scotia

WOLFFIA Horkel *ex* Schleid.

W. arrhiza (L.) Wimm.

LB P 1 George 1961: material from Zürich Botanic Garden

REFERENCES

Bischoff, H. W. and Bold, 1963. Phycological Studies IV. Univ. Texas Public. No. 6318

Borden, D., Whitt and Nanney, 1973. J. Protozool. 20(5) 693-700

Carr, N. G. and Whitton, 1973. The Biology of the Blue Green Algae. Blackwell

Chu, S. P., 1942. J. Ecol. 30 284-325

Corliss, J. O. 1971. Trans. Amer. Micros. Soc. 90(2) 243-252

Desikachary, T. V. 1972, Curr. Sci. 41(12) 445-447

Drouet, F. and Daily, 1956. Butler Univ. Bot. Stud. XII 1-222

Farooqui, P. B. 1974. Arch. f. Protistenk. 116 185-191

Fott, B. 1968. Acta Univ. Carol. Biol. 1967 223-240

Fott, B. 1971. Preslia (Praha) 43 289-303

Fott, B. Lockhead and Clemençon, 1975. Arch. Protistenk. 117 288-296

Fott, B. and Novakova, 1969. Studies in Phycology, Academia, Prague. 10-59

Gerloff, G. C., Fitzgerald and Skoog, 1950. in Culturing of Algae, ed. by Brunel, J., Prescott and Tiffany. Charles Kettering Foundation

Hoffman, L. R. and Hofmann, 1975. Can. J. Bot. 53 439-451

Hopkins, E. F. and Wann, 1926. Bot. Gaz. 81 353

Kantz, T. and Bold, 1969. Texas Univ. Publ, No. 6924

Kessler, E. 1974. Arch. Microbiol. 100 51-56

Komarek, J. 1970. Arch. Protistenk. 112 343-364

Kratz, W. A. and Myers, 1955. Am. J. Bot. 42 282-287

Morris, G. J. 1976. Arch. Microbiol. 107 57-62

Parke, M. and Raynes, 1964. J. mar. biol. Ass. 44 209-217

Pearsall, W. H. and Loose, 1937. Proc. R. Soc. London B 121 451-501

Pringsheim, E. G. 1928. Ber. deutsch. bot. Ges. XLVI 216-219

Pringsheim, E. G. 1948. New Phytol. 47(1) 52-87

Pringsheim, E. G. 1952. New Phytol. 51(1) 65-76

Provasoli, L., McLaughlin and Droop, 1957. Arch. Microbiol. 25 392-428

Robinson, D. G. and Preston, 1971. J. Exp. Bot. 22 635-643

Sorokin, C. and Krauss, 1958. Plant Physiol. 33 109-113

Stein, J. (Ed.) 1973. Handbook of Phycological Methods, Cambridge Univ. Press

Sueoka, N. 1960. Proc. Natl. Acad. Sci. 46 83-91

Venkataraman, G. S. 1969. The Cultivation of Algae; ICAR, New Delhi